Remote Sensing and Geographic Information System in Forestry

Remote Sensing and Geographic Information System in Forestry

Dr. Girish B. Shahapurmath
M.Sc. (Forestry) Ph.D.
Assistant Professor of Forest Management
Department of Natural Resource Management
College of Forestry, Sirsi (UK), Karnataka

Dr. A. G. Koppad
M.Sc. (Agri.) Ph.D. (Forestry)
Professor of Forest Engineering
Department of Natural Resource Management
College of Forestry, Sirsi (UK), Karnataka

SSPH

SATISH SERIAL PUBLISHING HOUSE
403, Express Tower, Commercial Complex
Azadpur, Delhi-110033 (India)
Phone : 011-27672852, Fax : 91-11-27672046
e-mail : info@satishserial.com, hkjain1975@yahoo.com
Website : www.satishserial.com

Published by :

SATISH SERIAL PUBLISHING HOUSE

403, Express Tower, Commercial Complex, Azadpur, Delhi-110033 (INDIA)
Phone : 011-27672852 Fax : 91-11-27672046
E-mail : info@satishserial.com, hkjain1975@yahoo.com

ISBN 978-93-88020-69-5

Composed, Designed & Printed in India

About the Authors

Dr. GIRISH B. SHAHAPURMATH, M.Sc. (Forestry) Ph.D.
Assistant Professor of Forest Management
Department of Natural Resource Management
College of Forestry, Sirsi-581 401, Karnataka
University of Agricultural Sciences, Dharwad, India
E-mail: girishbshahapur@gmail.com /
shahapurmathgb@uasd.in

- Specialization in Silviculture and Agroforestry, Forest Management, Remote Sensing and GIS.
- Post Graduate Diploma in Agricultural Extension and Management (PGDAEM) by MANAGAE, Hyderabad during 2014-15.
- Post Graduate Diploma Computer Applications (PGDCA) by Infotech Computer Education and Manipal Institute of Computer Education (MICE) during 2008-09.
- Total 16 years of experience in the field of forestry (Teaching, Research and Extension).
- Presently working as Assistant Professor of Forest Management in Department of Natural Resource Management at College of Forestry, SIRSI (University of Agricultural Sciences, Dharwad, Karnataka (India).
- Recipient of "Outstanding Scientist Award" in Forestry by The Society of Tropical Agriculture, New Delhi, India in 7th International Conference on Agriculture, Horticulture and Plant Sciences held at Shimla (H. P.) during June 28-29, 2018.
- Recipient of "Best Teacher Award-2016" for outstanding contribution in the field of Forestry on the occasion of International Conference on "Rural Livelihood Improvement by Enhancing Farmers' Income through Sustainable Innovative Agri and Allied Enterprises (RLSIAAe)" organized by the Society for Upliftment of Rural Economy, Varanasi (India) during 30th October to 1st November, 2018 held at Birla Institute of Technology, Patna.
- Recognized as a Life Member for the "Society for Upliftment of Rural Economy, Varanasi (Membership No. LM/245/2018).
- First prize for Best Poster Presentation on Multistoreyed Cropping of Teak-perennial vegetables-legume based Agroforestry system

Presented in National Symposium (Silver Jubilee) on "Agroforestry Knowledge for Sustainability, Climate Moderation and Challenges Ahead" held at National Research Centre for Agroforestry, Jhansi (UP) from 15-17 December, 2008.

- First Prize for Best Poster presentation on "Forest Law and Models" presented in "National Kaanoonu Utsava" Exhibition held at Sirsi from 07.06.2014 to 06.07.2014 organized by Karnataka High Court, Bengaluru and District Law Service, Karwar and Bar Association, Sirsi.
- Worked as Principal Investigator and CO-Principal Investigator for 10 Research Projects.
- Field of interest - Silviculture and Agroforestry, Forest Management and Policy, Sustainable Forest Management, Remote Sensing and Geographical Information System in Forestry.
- Developed the course materials as reference books and for competitive exams, such as ASRB, IFS, ACF, RFO, KPSC, etc., in the field of forestry for students' purpose.
- Attended more than 15 Trainings/Summer/Winter schools, short courses, Seminars, Workshops, Conferences and undergone Agricultural Massive Open Online Courses (Ag MOOC).
- Published more than 100 publications including Research papers in NAAS rated journal/Research notes/Papers presented and published in symposium/seminar/conference/workshop/Training/ international conference/Popular articles published in journals/ newspapers/Books/chapters in books.

Dr. A.G. KOPPAD M.Sc. (Agri). Ph.D. (Forestry)
Professor of Forest Engineering
Department of Natural Resource Management
College of Forestry, SIRSI-581 401, Karnataka
University of Agricultural Sciences, Dharwad, INDIA.
E-mail: koppadag@uasd.in / koppadag2007@rediffmail.com

- 24 years of experience in teaching in the field of Soil and Water Conservation Engineering, Forest hydrology, and Remote sensing and GIS field.
- Recipient of Incentive award for Adhoc project from University of Agricultural Sciences, Dharwad.
- Recipient of three best poster presentation awards.
- Visited Saskatoon (Canada) from 1/10/2010 to 15/11/2010 (45 days) and Kaulalampur (Malaysia) from 2 to 6th October, 2017.
- Research Areas of Interest - Soil and water conservation techniques and their impacts on forest species, Remote sensing and GIS applications in forestry.
- Organized two National seminars on medicinal plants and remote sensing and GIS field.
- Worked as Principal Investigator and Co- Principal Investigator for more than ten adhoc research projects from external funding agency.
- Published book on forest Engineering by ICAR for forestry graduates.
- Published more than 50 research papers in National and International Journals.

Hon'ble
Vice Chancellor

UNIVERSITY OF AGRICULTURAL SCIENCES, DHARWAD
KRISHI NAGAR, DHARWAD-581 005
KARNATAKA STATE, INDIA

Foreword

Remote sensing and GIS are complementary technologies that, when combined, enable improved monitoring, mapping, and management of forest resources. The information that supports forest management is stored primarily in the form of forest inventory databases within a GIS environment. A forest inventory is a survey of the location, composition, and distribution of forest resources. As one of the principal sources of forest management information, these databases support a wide range of management decisions from harvest plans to the development of long-term strategies. Historically, forest management inventories were primarily for timber management and focused on capturing area and volume by species. In the past decade, forest management responsibilities have broadened. As a result, inventory data requirements have expanded to include measures of non-harvest related characteristics, such as forest structure, wildlife habitat, biodiversity, and forest hydrology.

The entire forest inventory production cycle, from planning to map generation, can take several years. Except for the photo interpretation component, forest inventory production is largely a digital process. Operational level inventories, inventories based on both aerial photo interpretation and field-sampled measurements, provide location-specific information required for harvest planning. Forest management level inventories meet longer-term forest management planning objectives. Though these levels differ in detail, they both require information fundamentally based on forest inventory data. A forest management inventory generalizes complex forest resource attributes into mapping units useful for forest management. The types of attributes attached to individual mapping units, or polygons, might include stand species composition, density, height, age, and, more recently, new attributes, such as leaf area index. Much of the information collected for forest inventory is generated by interpretation of aerial photographs at photo scales of 1:10,000 to 1:20,000, depending on the level of detail required. Other remote sensing sources, such as airborne and satellite digital imagery has been valuable in updating forest attributes, such as disturbance, habitat, and biodiversity. In providing more frequent information updates, remotely sensed data can improve the quality of forest inventory databases, thereby improving the resource management activities they support.

The quality of photo interpreted data depends on the experience of the interpreters and the use of quality assurance procedures, such as interpreter calibration and field verification. Other factors can introduce inconsistencies that compromise the quality of forest inventory data. For example, there may be source data inconsistencies when aerial photography is acquired on different dates or in different weather conditions or inconsistencies in analysis when multiple contractors are used. The quality of the resulting data may vary significantly within a map area.

For example, information about disturbances related to fire and insects may be inconsistent within a map area because the aerial photography from which it was interpreted was acquired in different years. Similarly, inconsistencies may occur at the edge of neighbouring map sheets because data was collected in different years or was produced by different contractors. The use of remote sensing by forest managers has steadily increased, promoted in large part by better integration of imagery with GIS technology and databases, as well as implementations of the technology that better suit the information needs of forest managers.

A key development in remote sensing has been the increased availability of high-spatial- and high-spectral-resolution remotely sensed data from a wide range of sensors and platforms including photographic and digital cameras, video capture, and airborne and spaceborne multispectral sensors. Hyper spectral imagery promises to provide improved discrimination of forest cover and physiological attributes. Radar applications are being developed that penetrate the forest canopy to reveal characteristics of the forest floor. New technologies, such as LIDAR can provide estimates of forest biomass, height, and the vertical distribution of forest structure with unprecedented accuracy.

Textbooks on Remote Sensing and GIS in Forestry suitable for undergraduate and post graduate students are very few, particularly in Forestry Colleges in India. This textbook on "Remote Sensing and GIS in Forestry" by Dr. Girish B. Shahapurmath and Dr. A.G. Koppad is a timely and welcome contribution in the forestry field. The authors have a considerable experience in teaching, research and extension in forestry/ agroforestry and have presented the subject matter critically and systematically. This book will serve as a valuable source of information for the academicians in forestry institutes and also for the students for the preparation of competitive examinations in forestry discipline. I congratulate the authors for the hard work they put in, the systematic selection of topics and well documented and meticulous presentation.

I am hopeful that this text book will adequately meet the needs of students of forestry colleges and would be widely welcomed. I extend my best wishes to Dr. Girish B. Shahapurmath and Dr. A. G. Koppad for the success of their book.

May, 2019
Dharwad, Karnataka

Dr. M.B. CHETTI
(Vice Chancellor)
University of Agricultural Sciences, Dharwad
and Former Assistant Director General (HRD)
ICAR, New Delhi

Preface

Forestry, as a subject, is being taught to students of the Forest Colleges in different Agricultural Universities in India. This subject had been introduced as a subject for degree courses in most of the Indian Agricultural Universities. Indian Council of Agricultural Research (ICAR) provided an opportunity to start state-run forestry colleges in various State Agricultural Universities (SAUs) in the country. Earlier as per the recommendation of the National Commission on Agriculture, ICAR had started a course in 'Farm Forestry' during 1976 in all SAUs. There was a growing concern among the academicians and politicians alike that the students of forestry should have a greater exposure and access to natural forests. There has been increased public awareness of natural resource policy, including forest management. A key development in remote sensing has been the increased availability of high-spatial- and high-spectral-resolution remotely sensed data from a wide range of sensors and platforms including photographic and digital cameras, video capture, and airborne and spaceborne multispectral sensors. Hyper spectral imagery promises to provide improved discrimination of forest cover and physiological attributes. Radar applications are being developed that penetrate the forest canopy to reveal characteristics of the forest floor. New technologies, such as LIDAR can provide estimates of forest biomass, height, and the vertical distribution of forest structure with unprecedented accuracy. Therefore, this book has been written to meet the requirements of the students of the Forest Colleges as well as those of the students preparing for competitive examinations.

The book covers mostly the present syllabi of the Forest Colleges as well as those of the competitive examinations (Indian Forest Service and State Forest Services). I claim no originality in writing this book except for the arrangement and presentation of the subject matter. A large number of books, journals and proceedings of symposia and workshops have been consulted. A list of these references is given at the end of book as references/bibliography.

I, therefore, gratefully acknowledge the help taken from "Remote Sensing-Principles and Appliances" by B.C.Panda., "Remote Sensing and Forest Surveys" by M.K.Sharma., "Text book of Remote Sensing and Geographic Information System" by M. Anji Reddy., "Fundamentals of Geographic Information System' by Debashis Chakraborty and Rabi N Sahoo and other publications mentioned in the references.

I am greatly indebted to Dean (Forestry) for constant guidance and help during the time of writing the book and improving it substantially. I am equally grateful to all the faculty members of College of Forestry, Sirsi (UAS, Dharwad, Karnataka) and officers of Karnataka Forest Department for their valuable suggestions and help in writing the book as well as for reading through the chapters and making very valuable improvements.

Place: Sirsi, Karnataka
Date: 1st January, 2019

Dr. Girish B. Shahapurmath
Assistant Professor of Forest Management
Department of Natural Resource Management
College of Forestry, Sirsi (UAS, Dharwad)
Karnataka, India

Contents

Appendices

Chapter 1

Terms and Abbreviations used in Remote Sensing and GIS

Absorptance - A measure of the ability of a material to absorb EM energy at a specific wavelength.

Accuracy - The success in estimating the true value.

Active remote sensing - Remote sensing methods that provide their own source of electromagnetic radiation to illuminate the terrain. Radar is one example.

Additive colour process - A method for creating essentiall all colours through the addition (superposition) of light of the three colour primaries.

Additive primary colours - Blue, green, and red. Filters of these colours transmit the primary colour of the filter and absorb the other two colours.

Aerial photograph - A photograph of the Earth's surface taken (either directly downward or obliquely) from an airplane.

Aerial reconnaissance - The securing of information by aerial photography or by visual observation from the air.

Albedo (A) - Ratio of the amount of electromagnetic energy reflected by a surface to the amount of energy incident upon it.

Altimeter - An instrument which indicates the vertical distance above a specified datum plane.

Analog display - A form of data display in which values are shown in graphic form, such as curves. Differs from digital displays in which values are shown as arrays of numbers.

Analog - A form of display in which values are shown in graphic form, such as curves.

Analogue image - An image where the continuous variation in the property being sensed is represented by a continuos variation in image tone. In a photograph, this is achieved directly by the grains of photosensitive chemicals in the film; in an electronic scanner, the response in, say, millivolts is transformed to a display on a cathode - ray tube where it may be photographed.

Ancillary data - In remote sensing, secondary data pertaining to the area or classes of interest, such as topographic, demographic, or climatological data.

Angstrom - Unit of measurement = 10^{-8}cm.

Attenuation - A phyical process by which the flux density (or power, amplitude, intensity, illuminance) of a paralle beam of energy decreases with increasing distance from the source.

Azimuth - Geographic orientation of a line given as an angle measured in degrees clockwise from north.

Azimuth direction - In radar images, the direction in which the aircraft is heading. Also, called flight direction.

Azimuth resolution - In radar images, the spatial resolution in the azimuth direction.

Band - A wavelength interval in the electromagnetic spectrum. For example, in Landsat images the bands designate specific wavelength intervals at which images are acquired.

Binary - Numerical system using the base 2.

Bit-Contraction of binary digit, which in digital computing represents an exponent of the base 2.

Blackbody - An ideal substance that absorbs all the radiant energy incident on it and emits radiant energy at the maximum possible rate per unit area at each wavelength for any given temperature. No actual substance is a true blackbody, although some substances, such as lampblack, approach its properties.

Byte - A group of eight bits of digital data.

Calibration - Process of comparing an instrument's measurements with a standard.

Cartography - Map and chart construction.

CCD - Charge - coupled detector.

CCT - Computer - compatible tape.

Charge - coupled detector (CCD) - A device in which electron are stored at the surface of a semi-conductor.

Chlorosis - Yellowing of plant leaves resulting from an imbalance in the iron metabolism.

Colour composite image - Colour image prepared by projecting individual black and white multispectral images, each through a different colour filter. When the projected images are superposed, a colour composite image results.

Colour infrared film - Photographic film whose response extends to 0.9 μm; especially useful in distinguishing true vegetation and its condition from artificial types.

Colour ratio composite image - Colour composite image prepared by combining individual ratio images for a scene using a different colour for each ratio image.

Complementary colours - Two primary colours of light (one additive and the other subtractive) that produce white light when added together. Red and cyan are complimentary colours.

Data processing - Application of procedures mechanical, electrical, computer - whereby data are changed from one form to another.

Densitometer - Optical device for measuring the density of photographic transparencies.

Density of images - Measure of the opacity, or darkness, of a negative or positive transparency.

Depolarized - Refers to a change in polarization of a transmitted radar pulse as a result of various interactions with the terrain surface.

Depression angle (y) - In radar, the angle between the imaginary horizontal plane passing through the antenna and the line connecting the antenna and the target.

Digital display - A form of data display in which values are shown as arrays of numbers.

Digital image - An image where the property being measured has been converted from a continuous range of analogue values to a range expressed by a finite number of integers, usually recorded as binary codes from 0 to 255, or as one byte.

Digital image processing - Computer manipulation of the digital - number values of an image.

Digital number (DN) - Value assigned to a pixel in a digital image.

Digitization - Process of converting an analog display into a digital display.

Doppler principle - Describes the change in observed frequency that electromagnetic or other waves undergo as a result of the movement of the source of waves relative to the observer.

Electromagnetic radiation (EMR) - Energy propagated in the form of and advancing interaction between electric and magnetic fields. All electromagnetic radiation moves at the speed of light.

Electromagnetic spectrum - Continuous sequence of electromagnetic energy arranged according to wavelength or frequency.

Element - The smallest definable object of interest in the survey; it is a single item in a collection, population, or sample.

Emission - Process by which a body radiates electromagnetic energy. Emission is determined by kinetic temperature and emissivity.

Emissivity (e) - Ratio of radiant flux from a body to that from a blackbody at the same kinetic temperature and emissivity.

Emittance - A term for the radiant flux of energy per unit area emitted by a body.

Emulsion - Suspension of photosensitive silver halide grains in gelatin that constitutes the image - forming layer on photographic film.

EROS - Earth Resources Observation System.

ERTS - Earth Resource Technology Satellite, now called Landsat.

False colour image - A colour image where parts of the non - visible EM spectrum are expressed as one or more of the red, green, and blue components, so that the colours produced by the Earth's surface do not correspond to normal visual experience. Also called a false - colour composite (FCC). The most commonly seen false - colour images display the very - near infrared as red, red as green, and green as blue.

False colour photograph - Another term for IR colour photograph.

Far infrared - Extends through the thermal or emissive infrared region (5.6-1000 μm).

Field of view - The solid angle through which an instrument is sensitive to radiation.

Film - Light - sensitive photographic emulsion and its base.

Filter, digital - Mathematical procedure for modifying values of numerical data.

Filter, optical - A material that, by absorption or reflection, selectivity modifies the radiation transmitted through an optical system.

Filtering - In analysis, the removal of certain spectral or spatial frequencies to highlight features in the remaining image.

Focal length - In cameras, the distance from the optical centre of the lens to the plane at which the image of a very distant object is brought into focus.

Focus - The point at which the rays from a point source of light re-unite and cross after passing through a camera lens; the plane in which the sharpest image is formed.

GCP - Ground - control point: a geographic feature of known location that is recognizable on an image and can be used to determine geometrical corrections.

Geographic Information System (GIS) - A data handling and analysis system based on sets of data distributed spatially in two dimensions. The data sets may be map oriented, when they comprise qualitative attributes of an area recorded as lines, points, and areas often in vector format, or image oriented, when the data are quantitative attributes referring to cells in a rectangular grid usually in raster format. It is also known as a geobased or geocoded information system.

Geoid - The figure or shape of the Earth as generalized by extending mean sea level continuously through the continents.

Geometric correction - Image processing procedure that corrects spatial distortions in an image.

Geospatial - The spatial or dimensional characteristics of geographically defined features or conditions.

Geostationary - Refers to satellites travelling at the angular velocity at which the earth rotates; as a result, they remain above the same point on the earth at all times.

Geostationary Operational Environmental Satellite - a NOAA satellite that acquires visible and thermal IR images for meteorologic purposes.

Geostationary orbit - An orbit at 36,000 km in the direction of the Earth's rotation, which matches speed so that a satellite remains over a fixed point on the Earth's surface.

Geothermal - Refers to heat from sources within the earth.

Giga - A prefix meaning 10^9.

Goddard Space Flight Centre (GSFC) - The NASA facility at Greenbelt, Maryland, that is also a Landsat ground receiving station.

GMT - Greenwich mean time. This international 24h system is used to designate the time at which Landsat images are acquired.

GOES - Geostationary Operational Environmental Satellite.

Ground - control point - A geographic feature of known location that is recognizable on images and can be used to determine geometric corrections.

Ground range - On radar images, the distance from the ground track to an object.

Ground receiving station - Facility that records data transmitted by a satellite, such as Landsat.

Ground swath - Width of the strip of terrain that is imaged by a scanner system.

Ground truth - Jargon coined for data, information, and observations obtained on the surface or subsurface about features or classes, used to aid in interpretation of remotely sensed data; ground data is a preferred term.

IFOV - Instantaneous field of view; that recorded from a small area in a very brief interval of time.

Image - Pictorial representation of a scene recorded by a remote sensing system. Although image is a general term, it is commonly restricted to representations acquired by non - photographic methods.

Image dissection - The breaking down of a continuous scene into discrete spatial elements, either by the receptors on the retina, or in the process of capturing the image artificially.

Image enhancement - Any of a group of operations (e.g., contrast stretch, spatial filtering) that improve the detectability and presentation of targets or categories in an image.

Image processing - Encompasses all the various operations that can be applied to producing the final photographic or computer based image rendition.

Image striping - A defect produced in line scanner and pushbroom imaging devices produced by the non-uniform response of a single detector, or amongst a bank of detectors. In a linescan image, the stripes are perpendicular to flight direction, but parallel to it in a pushbroom image.

Image swath - See ground swath.

Instantaneous field of view (IFOV) - Solid angle through which a detector is sensitive to radiation. In a scanning system, the solid angle subtended by the detector when the scanning motion is stopped.

Interpretation - The process in which a person extracts information from an image.

IR - Infrared region of the electromagnetic spectrum that includes wavelengths from 0.7 μm to 1 mm.

IR color photograph - Colour photograph in which the red imaging layer is sensitive to photographic IR wavelengths, the green imaging layer is sensitive to red light, and the blue imaging layer is sensitive to green light. Also, known as camouflage detection photographs and false colour photographs.

Irradiance - The measure, in power units, of radiant flux incident on a surface; it has the dimensionality of energy per unit time (e.g., Watts)

Kirchhoff's Law - The radiation law which states that at a given temperature the ratio of the emissivity to the absorptivity for a given wavelength is the same for all bodies and is equal to the emissivity of an ideal blackbody at that temperature and wavelength.

Lambertian surface - An ideal, perfectly diffusing surface, which reflects energy equally in all directions.

Landsat - A series of unnamed earth - orbiting NASA satellites that acquire multispectral images in various visible and IR bands.

Laplacian filter - A form of non-directional digital filter.

Large scale - Aerial photography with a representative fraction of 1:500 to 1:10,000; maps with a RF (scale) greater than 1:1,00,000.

Laser - Light artificially stimulated electromagnetic radiation: a beam of coherent radiation with a single wavelength.

Lidar - Light intensity detection and ranging, which uses lasers to stimulate fluorescence in various compounds and to measure distances to reflecting surfaces.

Map projection - A systematic drawing of lines on a plane surface to represent the parallels or latitude and the meridians or longitude of the Earth.

Mid - infrared (MIR) - The range of EM wavelengths from 2.5 to 5.6 μm.

Mie scattering - The scattering of EM energy by particles in the atmosphere with comparable dimensions to the wavelength involved.

Modular Optoelectric Multispectral Scanner (MOMS) - An along track scanner carried on the Space Shuttle that recorded two bands of data.

MOMS - Modular Optoelectric Multispectral Scanner.

NASA - National Aeronautical and Space Administration.

Near infrared (NIR) - The shorter wavelength range of the infrared region of the EM spectrum, from 0.7 to 2.5 μm. It is often divided into very - near infrared (VNIR) covering the range accessible to photographic emulsions (0.7 to 1.0 μm), and the short - wavelength infrared (SWIR) covering the remainder of the NIR atmospheric window from 1.0 to 2.5 μm.

Negative photograph - Photograph on film or paper in which the relationship between bright and dark tones is the reverse of that of the features on the terrain.

Noise - Random or repetitive events that obscure or interfere with the desired information.

Non - selective scattering - The scattering of EM energy by particles in the atmosphere which are much larger than the wavelengths of the energy, and which causes all wavelengths to be scattered equally.

Oblique photograph - Photograph acquired with the camera intentionally directed at some angle between horizontal and vertical orientations.

Orbit - Path of a satellite around a body, such as the earth, under the influence of gravity.

Orthophotograph - A vertical aerial photograph from which the distortions due to varying elevation, tilt, and surface topography have been removed, so that it represents every object as if viewed directly from above.

Orthophotoscope - An optical - electronic device which converts a normal vertical aerial photograph to an orthophotograph.

Overlay - A transparent sheet giving information to supplement that shown on maps; a tracing of selected details on a photograph.

Overlap - Extent to which adjacent images or photographs cover the same terrain, expressed as a percentage.

Panchromatic film - Black and white film that is sensitive to all visible wavelengths.

Passive remote sensing - Remote sensing of energy naturally reflected or radiated from the terrain.

Path - and - row index - System for locating Landsat MSS and TM images.

Photograph - Representation of targets on film that results from the action of light on silver halide grains in the film's emulsion.

Photometry - The study of the measurement of intensity of light.

Photon - Minimum discrete quantity of radiant energy.

Picture - Representation of a scene by a photographic positive print or transparency, made from a negative, produced by EMR within or outside the visible range, converted to the visible by an optical-mechanical or electronic device.

Picture element - In a digitized image, the area on the ground represented by each digital number. Commonly contracted to pixel.

Pixel - Contraction of picture element; the spatial variable defines the apparent size of the resolution cell and the spectral variable defines the intentsity of the spectral response (at specified wavelength[s]) for that cell.

Planck's Law - An expression for the variation of emittance of a blackbody at a particular temperature as a function of wavelength.

Polarization - The direction of orientation in which the electrical field vector of electromagnetic radiation vibrates.

Precision - A (statistical) measure of the dispersion of the values observed when measuring a characteristic of elements in a population; a clustering of values about their average.

Primary colours - A set of three colours that in various combinations will produce the full range of colours in the visible spectrum. There are two sets of primary colours, additive and subtractive.

Principal - component (PC) image - Digitally processed image produced by a transformation that recognizes maximum variance in multispectral images.

Probability Density Function (PDF) - A function indicating the relative frequency with which any measurement may be expected to occur. In remote sensing it is represented by the histogram of DN in one band for a scene.

Radar - Acronym for radio detection and ranging. Radar is an active form of remote sensing that operates in the microwave and radio wavelength regions.

Radiation - Propagation of energy in the form of electromagnetic waves.

Radiometer - Device for quantitatively measuring radiant energy, especially thermal radiation.

Raster - The scanned and illuminated area of a video display, produced by a modulated beam of electrons sweeping the phosphorescent screen line by line from top to bottom at a regular rate of repetition.

Raster format - A means of representing spatial data in the form of a grid of DN, each line of which can be used to modulate the lines of a video raster.

Raster pattern - Pattern of horizontal lines swept by an electron beam across the face of a CRT that constitute the image display.

Rayleigh criterion - In radar, the relationship between surface roughness, depression angle, and wavelength that determines whether a surface will respond in a rough or smooth fashion to the radar pulse.

Rayleigh scattering - Selective scattering of light in the atmosphere by particle that are small compared with the wavelength of light.

Rectification - The process of projecting or transforming a tilted or oblique photograph onto a horizontal reference plane, the angular relation between the photograph and the plane being determined by ground reconnaissance.

Reflectance - Ratio of the radiant energy reflected by a body to the energy incident on it. Spectral reflectance is the reflectance measured within a specific wavelength interval.

Registration - Process of superposing two or more images or photographs so that equivalent geographic points coincide.

Relief displacement - Geometric distortion on vertical aerial photographs. The tops of objects appear in the photograph to be radially displaced from their bases outward from the photograph's centre point.

Remote sensing - Collection and interpretation of information about an object without being in physical contact with the object.

Representative fraction (RF) - The relation between map or photo distance and ground distance, expressed as a fraction or a ratio; also called scale.

Resolution - Ability to separate closely spaced objects on an image or photograph. Resolution is commonly expressed as the most closely spaced line-pairs per unit distance that can be distinguished. Also, called spatial resolution.

Scattering - Multiple reflections of electromagnetic waves by particles in the atmosphere or from surfaces.

Signature - Any characteristic or series of characteristics by which a material or object may be recognized in an image, photo, or dat set.

Sonar - Acronym for sound navigation ranging. Sonar is an active form of remote sensing that employs sonic energy to image the sea-floor.

Spectral reflectance - Reflectance of electromagnetic energy at specified wavelength intervals.

Spectral signature - Quantitative measurement of the properties of an object or material over one to several wavelength intervals; the diagnostic spectral response for that entity.

Spectral vegetation index - An index of relative amount and vigor of vegetation. The index is calculated from two spectral bands of AVHRR imagery.

Spectrometer - Device for measuring intensity of radiation absorbed or reflected by a material as a function of wavelength.

Spectroradiometer - A device which measures the energy reflected or radiated by materials in narrow EM wavebands.

Spectrum - Continuous sequence of electromagnetic energy arranged according to wavelength or frequency.

Stefan - Boltzmann constant - 5.68 x $10^{\wedge 12}$ W.cm $^{-2}$. K^{-4}; appropriate to Law of same name stating that the amount of energy radiated per unit time from a unit surface of an ideal blackbody is proportional to the fourth power of the absolute temperature (in units of K) of that body.

Stereopsis - The ability for objects to be perceived in three dimensions as a result of the parallax differences produced by the eye base.

Stereoscope - Binocular optical device for viewing overlapping images or diagrams. The left eye sees only the left image, and the right eye sees only the right image.

Supervised classification - Digital information extraction technique in which the operator provides training site information that the computer uses to assign pixels to categories.

Synthetic - aperture radar (SAR) - Radar system in which high azimuth resolution is achieved by storing and processing data on the Doppler shift of multiple return pulses in such a way as to give the effect of a much longer antenna.

Swath width - The overal plane angle or linear ground distance covered by a scanner in the across - track dimension.

Synchronous satellite - An equatorial west to east satellite orbiting the Earth at 34,900 km, at which altitude it makes one revolution in 24 hours synchronous with the Earth's rotation.

Synoptic view - The ability to see or otherwise measure widely dispersed areas at the sme time and under the same conditions.

Thematic map - A map designed to demonstrate particular features or concepts.

Thematic Mapper (TM) - A cross-track scanner deployed on Landsat that records seven bands of data from the visible through the thermal IR regions.

Thermal band - Generally, refers to thermal IR intervals associated with atmospheric windows, at 3-5 and 8-14 μm wavelengths.

Tone - Each distinguishable shade of grey from white to black on an image.

Transmittance - The ratio of the radint energy transmitted through a body to that incident upon it.

Unsupervised classification - Digital information extraction technique in which the computer assigns pixels to categories with no instructions from the operator.

UV - Ultraviolet region of the electromagnetic spectrum ranging in wavelengths from 0.01 to 0.4m.

Vector format - The expression of points, lines, and areas on a map by digitized Cartesian coordinates, directions, and values.

Wien's displacement law - Describes the shift of the radiant power peak to shorter wavelengths as temperature increases.

Zenith - The point in the celestial sphere that is exactly overhead; opposed to nadir.

Definition of Remote Sensing Terms

A discussion of remote sensing platforms, sensors and derivative products requires a vocabulary describing fundamental characteristics of the data. In this unit, accuracy, precision and resolution will be defined specifically for use in remote sensing applications.

Accuracy: The closeness of an estimated, measured, or computed value to a standard, accepted, or true value of a particular quantity.

Relative Accuracy: An evaluation of the amount of error in determining the location of one point or feature with respect to another. For example, the difference in elevation between two points on the earth's surface may be measured very accurately, but the stated elevations of the individual points with respect to the reference datum could contain a large error. In this case, the relative accuracy of the point elevations is high, but the absolute accuracy is low.

Precision: A statistical measure of the tendency for independent, repeated measurements of a value to produce the same result. A measurement can be highly repeatable, therefore, very precise, but inaccurate if the measuring instrument is not calibrated correctly. The same error would be repeated precisely in every measurement, but none of the measurements would be accurate.

Spectral Resolution: Describes the way an optical sensor responds to various wavelengths of light. High spectral resolution means that the sensor distinguishes between very narrow bands of wavelength; low spectral resolution means the sensor records the energy in a wide band of wavelengths as a single measurement.

Radiometric Resolution: Refers to the ability of a sensor to detect differences in energy magnitude. Sensors with low radiometric resolution are able to detect only relatively large differences in the amount of energy received; sensors with high radiometric resolution are able to detect relatively small differences.

Temporal Resolution: The frequency at which data are captured for a specific place on the earth. The more frequently they are captured, the better or finer is the temporal resolution. Temporal resolution is relevant when using imagery or elevations datasets captured successively over time to detect changes to the landscape.

Spatial Resolution: One of the more frequently misused terms in mapping specifications. According to the *Glossary of Mapping Sciences* (ASCE, ACSM, ASPRS, 1994), it is "a measure of the finest detail distinguishable in an image." While the distinguishable detail is dependent on the size of the image pixel, the size of an object that can be seen in an image and the size of a single pixel in an image are different.

Ground Sample Distance (GSD): The size of a pixel projected to the ground surface, as reported in linear units per pixel; for example, a USGS Digital Orthophoto Quarter Quad (DOQQ) has a 1-metre GSD because each pixel corresponds to a 1 metre by 1 metre square area on the ground. It is common to see the term "spatial resolution" used incorrectly in place of Ground Sample Distance (GSD). GSD is what many people (and remote sensing product vendors) actually mean when they talk about the "spatial resolution" of an image; GSD is the correct term.

Ground sample distance and spatial resolution are also frequently confused with spatial accuracy (location of that pixel with reference to a mapping datum). To say that the size of a pixel in an image is 1 foot in ground units is not the same as saying that the ground coordinates of that pixel are accurate to within 1 foot of their "true value." Thus, an image can have very high spatial resolution and very low spatial accuracy, or vice versa.

When using the term in reference to an elevation model, GSD describes the actual or nominal spacing between ground elevation samples or measurements.

Post Spacing: Describes the ground distance interval of cells in a uniform elevation grid. For example, a USGS Digital Elevation Model (DEM), one of the standard products in the National Elevation Dataset (NED) has 30-metre post spacing; the term is synonymous with grid spacing. It is not exactly the same as GSD in reference to elevation models. GSD refers to the spacing of the actual measurements, whereas post spacing refers to the grid interval of the interpolated product generated from those measurements.

Forests and Remote Sensing Overview

Remote sensing allows for efficient management of forest resources. These resources include wood, forage, water, wildlife, recreation, timber management, maintenance and improvement of existing forest stands and fire control. Satellite and remote sensing images, with airphoto interpretation, combined with forest classification maps, will help to determine the volume of timber harvested from an individual tree or stand of trees.

Remote sensing images will detect signs of damage of forests by fire, insects or disease. Airphotos will also provide information about range and wildlife habitat management, in outdoor recreation surveys and in estimating the volumes of the standing trees. Seasonal differences should be accounted for when viewing airphoto/satellite images. Individual tree species will be identifiable through shape, size, pattern, shadow, tone and texture. Measurements include tree height/stand height, tree crown diameter, density of stocking and stand area. Compact Airborne Spectrographic Imager (CASI) provides images and has been used to calculate the Leaf Area Index for a study site in the Kananaskis Valley.

The emphasis is placed on forest inventory techniques and detection of plants on large-scale photographs, as well as looking for signs of damage by fire, insects, or disease.

Global Positioning Systems (GPS) assist in the management of forests. As the size and physical terrain of forests make it difficult to survey, GPS makes it easier, faster and more accurate to create land surveys (within 1% of land survey accuracies using DGPS). Also, GPS allows for measurement of timber of interest to forestry companies and conservation groups. Forest fire management is enhanced by the power of Differential Global Positioning Systems (DGPS). For example, a helicopter with DGPS gear can fly along the perimeter of a fire and download an accurate map of the fire's size. This provides useful information to destroy the fire in the right places with the right number of firefighters. DGPS was designed and used by the U.S. Forest Service.

❑❑❑

Chapter 2

Remote Sensing - Introduction, Definition, Brief History, Principles, Steps, Merits, Limitations, Classification

Introduction

Remote sensing (RS), also called earth observation, refers to obtaining information about objects or areas at the Earth's surface without being in direct contact with the object or area. Humans accomplish this task with aid of eyes or by the sense of smell or hearing; so, remote sensing is day-to-day business for people. Reading the newspaper, watching cars driving in front of you are all remote sensing activities. Most sensing devices record information about an object by measuring an object's transmission of electromagnetic energy from reflecting and radiating surfaces.

Remote sensing techniques allow taking images of the earth surface in various wavelength region of the electromagnetic spectrum (EMS). One of the major characteristics of a remotely sensed image is the wavelength region it represents in the EMS. Some of the images represent reflected solar radiation in the visible and the near infrared regions of the electromagnetic spectrum, others are the measurements of the energy emitted by the earth surface itself i.e. in the thermal infrared wavelength region. The energy measured in the microwave region is the measure of relative return from the earth's surface, where the energy is transmitted from the vehicle itself. This is known as ***active remote sensing,*** since the energy source is provided by the remote sensing platform. Whereas, the systems where the remote sensing measurements depend upon the external energy source, such as sun are referred to as *passive remote sensing* systems.

Historic Overview

In 1859, Gaspard Tournachon took an oblique photograph of a small village near Paris from a balloon. With this picture the era of earth observation and remote sensing had started. His example was soon followed by other people all over the world. During the Civil War in the United States aerial photography from balloons played an important role to reveal the defence positions in Virginia (Colwell, 1983). Likewise other scientific and technical developments, this Civil Wartime in the United States speeded up the development of photography, lenses and applied airborne use of this technology. Table 1 shows a few important dates in the development of remote sensing.

The next period of fast development took place in Europe and not in the United States. It was during World War I that aeroplanes were used on a large scale for photoreconnaissance. Aircraft proved to be more reliable and more stable platforms for earth observation than balloons. In the period between World War I and World War II, a start was made with the civilian use of aerial photos. Application fields of airborne photos included at that time geology, forestry, agriculture and cartography. These developments lead to much improved cameras, films and interpretation equipment. **The most important developments of aerial photography and photo interpretation took place during World War II**. During this time-span the development of other imaging systems, such as near-infrared photography; thermal sensing and radar took place. Near-infrared photography and thermal-infrared proved very valuable to separate real vegetation from camouflage. The first successful airborne imaging radar was not used for civilian purposes but proved valuable for nighttime bombing. As such the system was called by the military 'plan position indicator' and was developed in Great Britain in 1941.

Table 1: Milestones in the History of Remote Sensing

1800	Discovery of Infrared by Sir W. Herschel
1839	Beginning of Practice of Photography
1847	Infrared Spectrum Shown by J.B.L. Foucault
1859	Photography from Balloons
1873	Theory of Electromagnetic Spectrum by J.C. Maxwell
1909	Photography from Airplanes
1916	World War I: Aerial Reconnaissance
1935	Development of Radar in Germany
1940	WW II: Applications of Non-Visible Part of EMS
1950	Military Research and Development
1959	First Space Photograph of the Earth (Explorer-6)
1960	First TIROS Meteorological Satellite Launched
1970	Skylab Remote Sensing Observations from Space
1972	Launch Landsat-1 (ERTS-1) : MSS Sensor
1972	Rapid Advances in Digital Image Processing
1982	Launch of Landsat -4 : New Generation of Landsat Sensors: TM
1986	French Commercial Earth Observation Satellite SPOT
1986	Development Hyperspectral Sensors
1990	Development High Resolution Space borne Systems First Commercial Developments in Remote Sensing
1998	Towards Cheap One-Goal Satellite Missions
1999	Launch EOS : NASA Earth Observing Mission
1999	Launch of IKONOS, very high spatial resolution sensor

After the wars in the 1950s remote sensing systems continued to evolve from the systems developed for the war effort. **Colour infrared (CIR)** photography was found to be of great use for the plant sciences. In 1956, Colwell conducted experiments on the use of CIR for the classification and recognition of vegetation types and the detection of diseased and damaged or stressed vegetation. It was also in the 1950s that significant progress in radar technology was achieved.

Definitions

"Remote sensing is the science (and to some extent, art) of acquiring information about the Earth's surface without actually being in contact with it. This is done by sensing and recording reflected or emitted energy and processing, analyzing, and applying that information."

Remote Sensing : This is the collection of information relating to objects without being in physical contact with them. Thus, our eyes and ears are remote sensors, and the same is true for cameras and microphones and for many instruments used for all kinds of applications.

OR

Remote sensing : This is the process of acquiring data/ information about objects/substances not in direct contact with the sensor, by gathering its inputs using electromagnetic radiation or acoustical waves that emanate from the targets of interest. An aerial photograph is a common example of a remotely sensed (by camera and film, or now digital) product.

Types of Remote Sensing

The sun is a source of energy or radiation, which provides a very convenient source of energy for remote sensing. The sun's energy is either **reflected**, as it is for visible wavelengths, or absorbed and then **reemitted**, as it is for thermal infrared wavelengths.

There are two main types of remote sensing: **Passive remote sensing and Active remote sensing.**

1. **Passive remote sensing :** Passive sensors detect natural radiation that is emitted or reflected by the object or surrounding area being observed. Reflected sunlight is the most **Passive Active** common source of radiation measured by passive sensors. Examples of passive remote sensors include film photography, infrared, and radiometers.
2. **Active remote sensing :** Active remote sensing, on the other hand, emits energy in order to scan objects and areas, whereupon a sensor then detects and measures the radiation that is reflected or backscattered from the target. RADAR is an example of active remote sensing where the time delay between emission and return is measured, establishing the location, height, speeds and direction of an object.

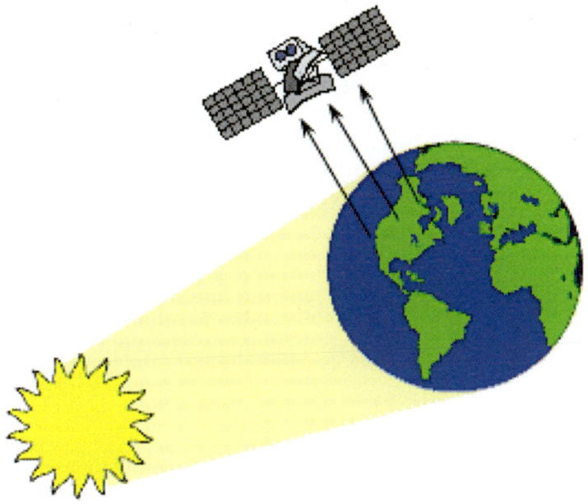

1. Passive remote sensing

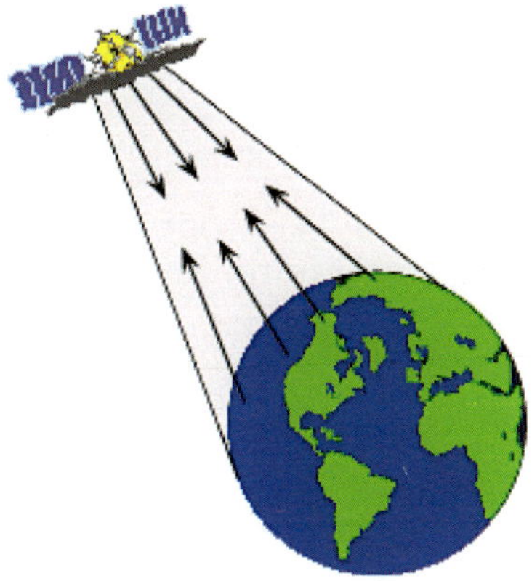

2. Active remote sensing

Overview

Remote sensing makes it possible to collect data on dangerous or inaccessible areas. Remote sensing applications include monitoring deforestation in areas, such as the Amazon Basin, the effects of climate change on glaciers and Arctic and Antarctic regions, and depth sounding of coastal and ocean depths. Military collection during the cold war made use of stand-off collection of data about dangerous border areas. Remote sensing also replaces costly and slow data collection on the ground, ensuring in the process that areas or objects are not disturbed.

Applications of Remote Sensing

There are probably hundreds of applications - these are typical:

Meteorology - Study of atmospheric temperature, pressure, water vapour, and wind velocity.

Oceanography: Measuring sea surface temperature, mapping ocean currents, and wave energy spectra and depth sounding of coastal and ocean depths.

Glaciology- Measuring ice cap volumes, ice stream velocity, and sea ice distribution. **(Glacial)**

Geology- Identification of rock type, mapping faults and structure.

Geodesy- Measuring the figure of the Earth and its gravity field.

Topography and cartography - Improving digital elevation models.

Agriculture - Monitoring the biomass of land vegetation

Forest - Monitoring the health of crops, mapping soil moisture

Botany - Forecasting crop yields.

Hydrology - Assessing water resources from snow, rainfall and underground aquifers.

Disaster warning and assessment - Monitoring of floods and landslides, monitoring volcanic activity, assessing damage zones from natural disasters.

Planning applications - Mapping ecological zones, monitoring deforestation, monitoring urban land use.

Oil and mineral exploration - Locating natural oil seeps and slicks, mapping geological structures, monitoring oil field subsidence.

Military - Developing precise maps for planning, monitoring military infrastructure, monitoring ship and troop movements

Urban - Determining the status of a growing crop

Climate - The effects of climate change on glaciers and Arctic and Antarctic regions

Sea - Monitoring the extent of flooding

Rock - Recognizing rock types

Space program - This is the backbone of the space program

Seismology - As a premonition.

Principles and Process of Remote Sensing

Principles of Remote Sensing

Detection and discrimination of objects or surface features means detecting and recording of radiant energy reflected or emitted by objects or surface material (Fig. 1). Different objects return different amount of energy in different bands of the electromagnetic spectrum, incident upon it. This depends on the property of material (structural, chemical, and physical), surface roughness, angle of incidence, intensity, and wavelength of radiant energy. The Remote Sensing is basically a multi-disciplinary science which includes a combination of various disciplines, such as optics, spectroscopy, photography, computer, electronics and telecommunication, satellite launching, etc. All these technologies are integrated to act as one complete system in itself, known as Remote Sensing System. There are a number of stages in a Remote Sensing process, and each of them is important for successful operation.

Stages in Remote Sensing

- Emission of electromagnetic radiation, or **EMR** (sun/self-emission)
- Transmission of energy from the source to the surface of the earth, as well as absorption and scattering
- Interaction of **EMR** with the earth's surface: reflection and emission
- Transmission of energy from the surface to the remote sensor
- Sensor data output
- Data transmission, processing and analysis

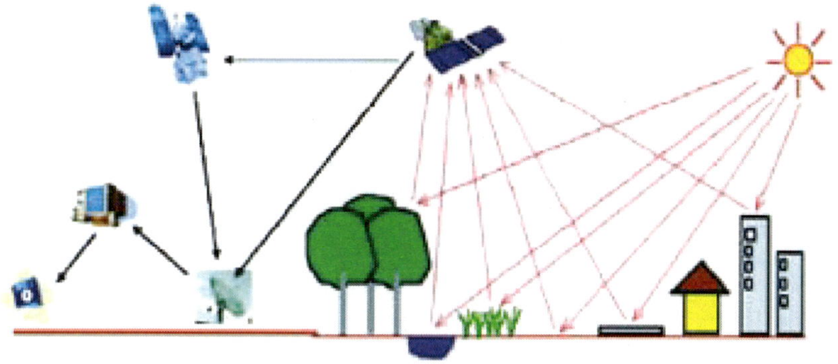

Figure 1: Remote Sensing process

Remote sensing actually done from satellites as Landsat or airplane or on the ground. To repeat the essence of the definition above, remote sensing uses instruments that house sensors to view the spectral, spatial and radiometric relations of observable objects and materials at a distance. Most sensing modes are based on sampling of photons corresponding frequency in the electromagnetic (EM) spectrum.

In much of remote sensing, the process involves an interaction between incident radiation and the targets of interest. This is exemplified by the use of imaging systems where the following **seven elements** are involved. Note, however, that remote sensing also involves the sensing of emitted energy and the use of non-emitted sensors.

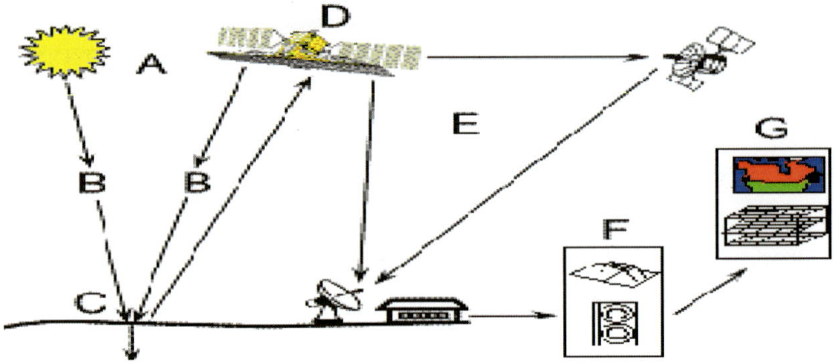

i) **Energy Source or Illumination (A)** - The first requirement for remote sensing is to have an energy source which illuminates or provides electromagnetic energy to the target of interest.

ii) **Radiation and the Atmosphere (B)** - As the energy travels from its source to the target, it will come in contact with and interact with the atmosphere it passes through. This interaction may take place a second time, as the energy travels from the target to the sensor.

iii) **Interaction with the Target (C)** - Once the energy makes its way to the target through the atmosphere, it interacts with the target depending on the properties of both the target and the radiation.

iv) **Recording of Energy by the Sensor (D)** - After the energy has been scattered by, or emitted from the target, we require a sensor (remote - not in contact with the target) to collect and record the electromagnetic radiation.

v) **Transmission, Reception, and Processing (E)** - The energy recorded by the sensor has to be transmitted, often in electronic form, to a receiving and processing station where the data are processed.

vi) **Interpretation and Analysis (F)** - The processed image is interpreted, visually and/or digitally or electronically, to extract information about the target, which was illuminated.

vii) **Application (G)** - The final element of the remote sensing process is achieved when we apply the information we have been able to extract from the imagery about the target in order to better understand it, reveal some new information, or assist in solving a particular problem.

❑❑❑

Chapter 3

Types of Remote Sensing Systems

Types of remote sensing system: There are nine different types of remote sensing system:

1. Visual Remote Sensing System
2. Optical Remote Sensing
3. Infrared Remote Sensing
4. Microwave Remote Sensing
5. Radar Remote Sensing
6. Satellite Remote Sensing
7. Airborne Remote Sensing
8. Acoustic and near-acoustic Remote Sensing
9. LIDAR Remote Sensing

1. Visual Remote Sensing System

The human visual system is an example of a remote sensing system in the general sense. **The sensors in this example are the two types of photosensitive cells, known as the cones and the rods, at the retina of the eyes**. The cones are responsible for colour vision. There are three types of cones, each being sensitive to one of the red, green, and blue regions of the visible spectrum. Thus, it is not coincidental that the modern computer display monitors make use of the same three primary colours to generate a multitude of colours for displaying colour images.

The cones are insensitive under low light illumination condition, when their jobs are taken over by the rods. The rods are sensitive only to the total light intensity. Hence, everything appears in shades of grey, when there is insufficient light. As the objects/events being observed are located far away from the eyes, the information needs a carrier to travel from the object to the eyes. In this case, the information carrier is the visible light, a part of the electromagnetic spectrum.

The objects reflect/scatter the ambient light falling onto them. Part of the scattered light is intercepted by the eyes, forming an image on the retina after passing through the optical system of the eyes. The signals generated at the retina are carried via the nerve fibres to the brain, the central processing unit (CPU) of the visual system. These signals are processed and interpreted at the brain, with the aid of previous experiences. The visual system is an example of a "***Passive Remote Sensing***" system which depends on an external source of energy to operate. We all know that this system won't work in darkness.

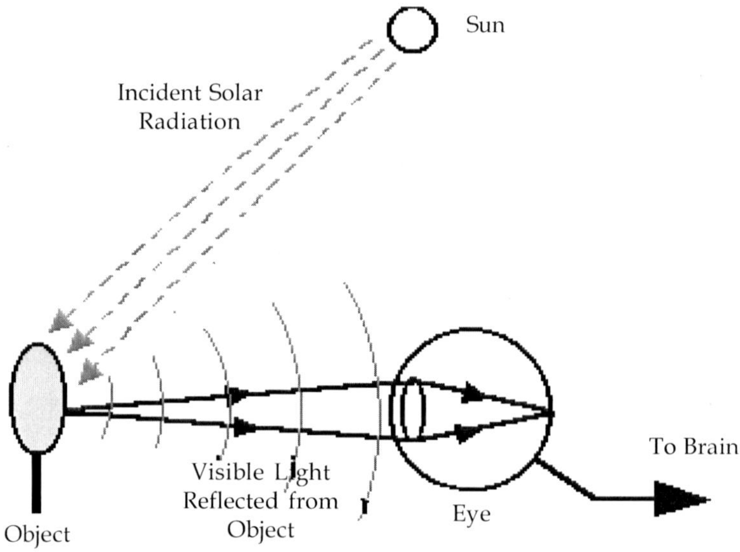

Fig. Visual remote sensing system

2. Optical Remote Sensing

In Optical Remote Sensing, optical sensors detect solar radiation reflected or scattered from the earth, forming images resembling photographs taken by a camera high up in space. The wavelength region usually extends from the visible and near infrared VNIR to the short-wave infrared SWIR. Different materials, such as water, soil, vegetation, buildings and roads reflect visible and infrared light in different ways. They have different colours and brightness, when seen under the sun. The interpretations of optical images requires the knowledge of the ***spectral reflectance signatures*** of the various materials (natural or man-made) covering the surface of the earth.

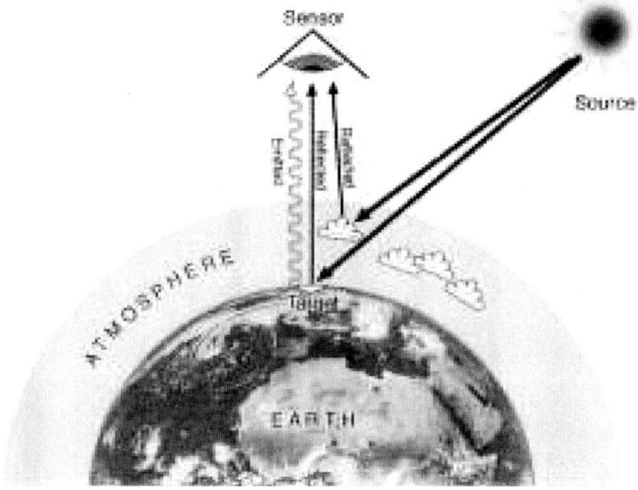

Fig. Optical Remote Sensing

3. Infrared Remote Sensing

Infrared remote sensing makes use of infrared sensors to detect infrared radiation emitted from the Earth's surface. The middle-wave infrared (MWIR) and long-wave infrared (LWIR) are within the thermal infrared region. These radiations are emitted from warm objects, such as the Earth's surface. They are used in satellite remote sensing for measurements of the earth's land and sea surface temperature. Thermal infrared remote sensing is also often used for detection of forest fires, volcanoes, oil fires.

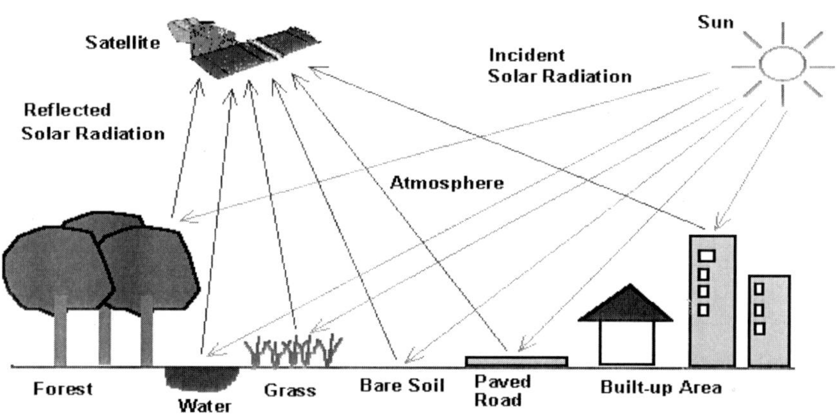

Fig. Infrared Remote Sensing

4. Microwave Remote Sensing

There are some remote sensing satellites which carry passive or active microwave sensors. The active sensors emit pulses of microwave radiation to illuminate the areas to be imaged. Images of the earth surface are formed by measuring the microwave energy scattered by the ground or sea back to the sensors. These satellites carry their own "flashlight" emitting microwaves to illuminate their targets. ***The images can, thus, be acquired day and night***. Microwaves have an additional advantage, as they can ***penetrate clouds***. Images can be acquired even when there are clouds covering the earth surface.

Fig. Microwave Remote Sensing

A microwave imaging system which can produce high resolution image of the Earth is the ***synthetic aperture radar*** **(SAR)**. Electromagnetic radiation in the microwave wavelength region is used in remote sensing to provide useful information about the Earth's atmosphere, land and ocean. When microwaves strike a surface, the proportion of energy scattered back to the sensor depends on many factors:

a) Physical factors, such as the ***dielectric constant*** of the surface materials which also depends strongly on the moisture content.

b) Geometric factors, such as surface ***roughness, slopes, orientation*** of the objects relative to the radar beam direction.

c) The types of ***landcover*** (soil, vegetation or man-made objects).

d) Microwave ***frequency, polarisation and incident angle.***

5. Radar Remote Sensing

Using radar, geographers can effectively map out the terrain of a territory. Radar works by sending out radio signals, and then waiting for them to bounce off the ground and return. By measuring the amount of time it takes for the signals to return, it is possible to create a very accurate topographic map. An important advantage to using radar is that it can ***penetrate thick clouds and moisture***. This allows scientists to accurately map areas, such as rain forests, which are otherwise too obscured by clouds and rain. Imaging radar systems are versatile sources of remotely sensed images, providing daynight, all-weather imaging capability.

Radar images are used to map landforms and geologic structure, soil types, vegetation and crops, and ice and oil slicks on the ocean surface.

Fig. Radar Remote Sensing

Synthetic Aperture Radar (SAR)

In synthetic aperture radar (SAR) imaging, microwave pulses are transmitted by an antenna towards the earth surface. The microwave energy scattered back to the spacecraft is measured. The SAR makes use of the radar principle to form an image by utilizing the time delay of the backscattered signals. In real aperture radar imaging, the ground resolution is limited by the size of the microwave beam sent out from the antenna.

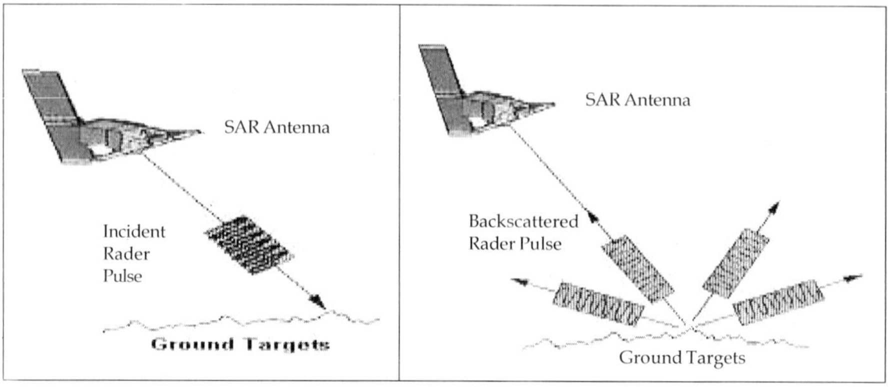

6. Satellite Remote Sensing

In this, you will see many remote sensing images acquired by earth observation satellites. These remote sensing satellites are equipped with sensors looking down to the earth. They are the **"eyes in the sky"** constantly observing the earth, as they go round in predictable orbits. Orbital platforms collect and transmit data from different parts of the electromagnetic spectrum, which in conjunction with larger scale aerial or ground-based sensing and analysis provides researchers with enough information to monitor trends. Other uses include different areas of the earth sciences, such as natural resource management, agricultural fields, such as land usage and conservation, and national security and overhead, ground-based and stand-off collection on border areas.

Fig. Satellite Remote Sensing

How Satellites Acquire Images?

Satellite sensors record the intensity of electromagnetic radiation (sunlight) reflected from the earth at different wavelengths. Energy that is not reflected by an object is absorbed. Each object has its own unique 'spectrum', some of which are shown in the diagram below. Remote sensing relies on the fact that particular features of the landscape such as bush, crop, salt-affected land and water reflect light differently in different wavelengths. Grass looks green, for example, because it reflects green light and absorbs other visible wavelengths. This can be seen as a peak in the green band in the reflectance spectrum for green grass above. The spectrum also shows that grass reflects even more strongly in the infrared part of the spectrum. While this can't be detected by the human eye, it can be detected by an infrared sensor. Instruments mounted on satellites detect and record the energy that has been reflected. The detectors are sensitive to particular ranges of wavelengths, called **'bands'**. The satellite systems are characterized by the bands at which they measure the reflected energy. The Landsat TM satellite, which provides the data used in this project, has bands at the blue, green and red wavelengths in the visible part of the spectrum and at three bands in the near and mid infrared part of the spectrum and one band in the thermal infrared part of the spectrum. The satellite detectors measure the intensity of the reflected energy and record it.

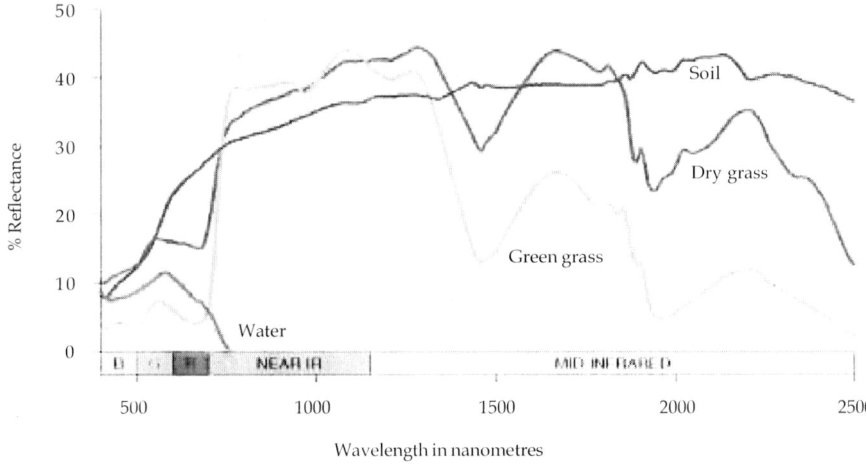

7. Airborne Remote Sensing

In airborne remote sensing, downward or sideward looking sensors are mounted on an aircraft to obtain images of the earth's surface. An advantage of airborne remote sensing, compared to satellite remote sensing, is the capability of offering very ***high spatial resolution*** images (20 cm or less). The disadvantages are ***low coverage area and high cost per unit area of ground coverage***. It is not cost-effective to map a large area using an airborne remote sensing system. Airborne remote sensing missions are often carried out as one-time operations, whereas earth observation satellites offer the possibility of continuous monitoring of the earth.

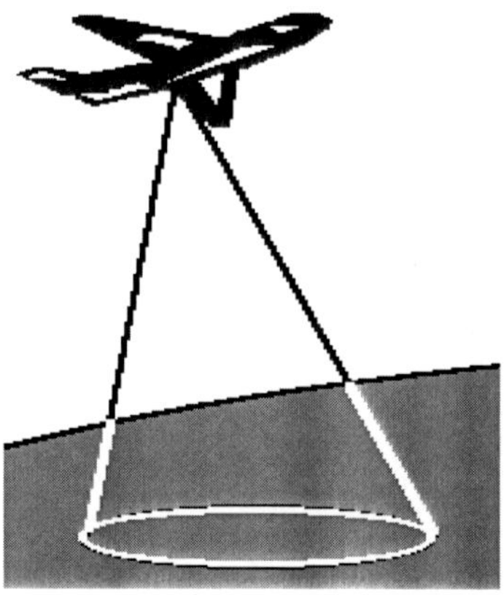

8. Acoustic and near-acoustic remote sensing

i) Sonar

   a) Passive sonar, listening for the sound made by another object (a vessel, a whale etc).

   b) Active sonar, emitting pulses of sounds and listening for echoes, used for detecting, ranging and measurements of underwater objects and terrain.

ii) **Seismograms** taken at different locations can locate and measure earthquakes (after they occur) by comparing the relative intensity and precise timing.

9. LIDAR Remote Sensing System

LIDAR or **LADAR** {**Light (Laser) Detection and Ranging**} is an optical remote sensing technology that can measure the distance to, or other properties of a target by illuminating the target with light, often using pulses from a laser. LIDAR technology has application in archaeology, geography, geology, geomorphology, seismology, forestry, remote sensing and atmospheric physics.

General description

LIDAR uses ultraviolet, visible, or near infrared light to image objects and can be used with a wide range of targets, including non-metallic objects, rocks, rain, chemical compounds, aerosols, clouds and even single molecules. A narrow laser beam can be used to map physical features with very high resolution.

LIDAR has been used extensively for atmospheric research and meteorology. Downward-looking LIDAR instruments fitted to aircraft and satellites are used for surveying and mapping.

Wavelengths in a range from about 10 micrometres to the UV (250 nm) are used to suit the target. Typically light is reflected via backscattering. Different types of scattering are used for different LIDAR applications, most common are Rayleigh scattering, Mie scattering and Raman scattering as well as fluorescence.

Suitable combinations of wavelengths can allow for remote mapping of atmospheric contents by looking for wavelength-dependent changes in the intensity of the returned signal.

Basic of LIDAR System

LIDAR system involves a laser range finder reflected by a rotating mirror (top). The laser is scanned around the scene being digitized, in one or two dimensions gathering distance measurements at specified angle intervals.

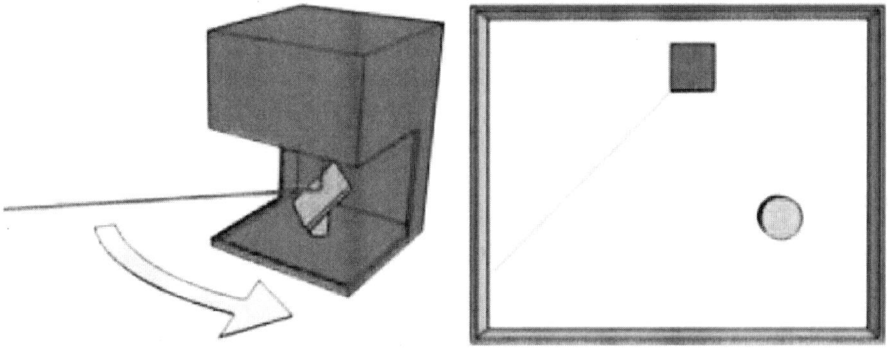

In general, there are two kinds of LIDAR detection schema

i) **Incoherent:** As direct energy detection (which is principally an amplitude measurement) and

ii) **Coherent detection:** (which is best for doppler, or phase sensitive measurements). Coherent systems generally use Optical detection which being more sensitive than direct detection allows them to operate a much lower power but at the expense of more complex transceiver requirements.

In both coherent and incoherent LIDAR, there are two types of pulse models:

a) Micropulse lidar systems and

b) High energy systems

a) **Micropulse systems** have developed as a result of the ever increasing amount of computer power available combined with advances in laser technology. They use considerably less energy in the laser, typically on the order of one microjoule, and are often "eye-safe," meaning they can be used without safety precautions.

b) **High-power systems** are common in atmospheric research, where they are widely used for measuring many atmospheric parameters: the height, layering and densities of clouds, cloud particle properties (extinction coefficient, backscatter coefficient, depolarization), temperature, pressure, wind, humidity, trace gas concentration (ozone, methane, nitrous oxide, etc.).

Major components of LIDAR system

1. **Laser:** 600–1000 nm lasers are most common for non-scientific applications. They are inexpensive but since they can be focused and easily absorbed by the eye the maximum power is limited by the need to make them eye-safe. A common alternative 1550 nm lasers are eye-safe at much higher power levels since this wavelength is not focused by the eye, but the detector technology is less advanced and so, these wavelengths are generally used at longer ranges and lower accuracies. They are also used for military applications as **1550 nm is not visible** in night vision goggles unlike the shorter 1000 nm infrared laser.

   Airborne topographic mapping LIDAR generally use 1064 nm diode pumped YAG lasers, while bathymetric systems generally use 532 nm frequency doubled diode pumped YAG lasers because 532 nm penetrates water with much less attenuation than does 1064 nm. Laser settings include the laser repetition rate (which controls the data collection speed). Pulse length is generally an attribute of the laser cavity length, the number of passes required through the gain material (YAG, YLF, etc.), and Q-switch speed. Better target resolution is achieved with shorter pulses, provided the LIDAR receiver detectors and electronics have sufficient bandwidth.

2. **Scanner and optics**: How fast images can be developed is also affected by the speed at which it can be scanned into the system. There are several options to scan the azimuth and elevation, including dual oscillating plane mirrors, a combination with a polygon mirror, a dual axis scanner. Optic choices affect the angular resolution and range that can be detected. A hole mirror or a beam splitter are options to collect a return signal. 3. **Photodetector and receiver electronics**: Two main photodetector technologies are used in LIDAR: solid state photodetectors, such as silicon avalanche photodiodes, or photomultipliers. The sensitivity of the receiver is another parameter that has to be balanced in a LIDAR design.

3. **Position and navigation systems**: LIDAR sensors that are mounted on mobile platforms, such as airplanes or satellites require instrumentation to determine the absolute position and orientation of the sensor. Such devices generally include a Global Positioning System receiver and an Inertial Measurement Unit (IMU).

Chapter 4

Electromagnetic Spectrum (EMS)

(Definition and descriptions, radiations laws of light, differential reflections, scattering by surfaces, atmospheric window, role in remote sensing).

Electromagnetic Waves

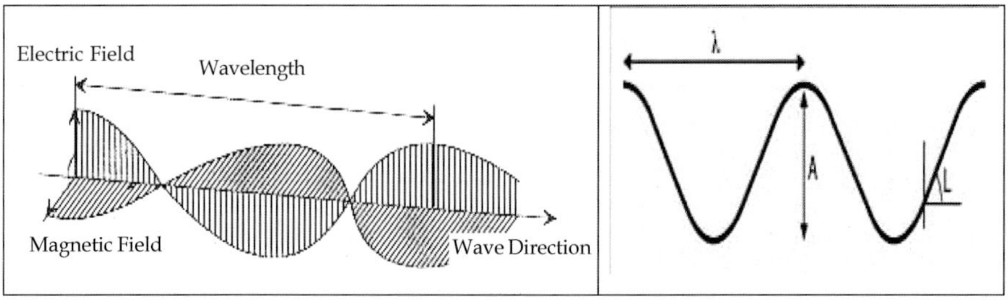

Electromagnetic waves are energy transported through space in the form of periodic disturbances of electric and magnetic fields. All electromagnetic waves travel through space at the same speed, **c = 2.99792458 x 10^8 m/s,** commonly known as the **speed of light**. An electromagnetic wave is characterized by a **frequency** and a **wavelength**.

Photons

According to quantum physics, the energy of an electromagnetic wave is quantized, i.e. it can only exist in discrete amount. The basic unit of energy for an electromagnetic wave is called a **photon**. The energy E of a photon is proportional to the wave frequency γ,

$E = h\gamma = hc/\lambda$, Where the constant h is the **Planck's Constant, h = 6.626 x 10-34 J s.**

The frequency (and hence, the wavelength) of an electromagnetic wave depends on its source. There is a wide range of frequency encountered in our physical world, ranging from the low frequency of the electric waves generated by the power transmission lines to the very high frequency of the gamma rays originating from the atomic nuclei. These wide frequency ranges of electromagnetic waves constitute the **Electromagnetic Spectrum**. The electromagnetic spectrum can be divided into several wavelength (frequency) regions, among which only a narrow band from about 400 to 700 nm is visible to the human eyes. Note that there is no sharp boundary between these regions. The boundaries shown in the figures are approximate and there are overlaps between two adjacent regions.

Wavelength units: 1mm = 1000, μm; 1μm = 1000 nm.

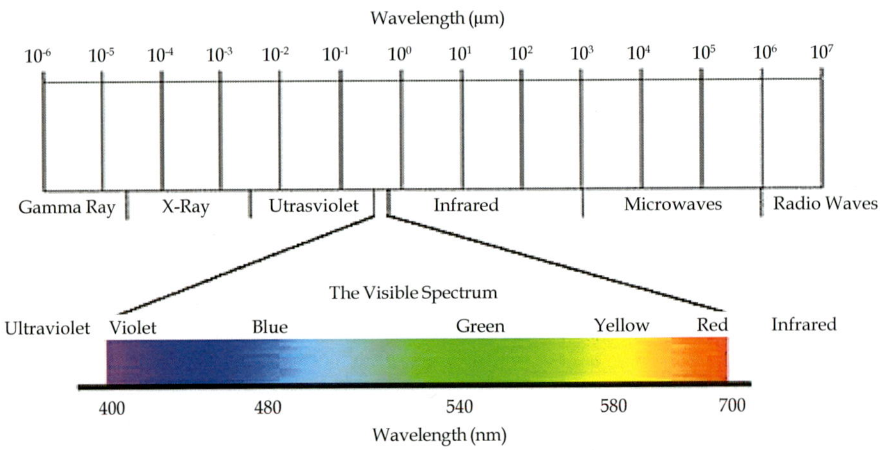

1. **Gamma Rays <0.30 nm:** This range is completely absorbed by the upper atmosphere and not available for remote sensing.
2. **X-Rays 0.03–30.0 nm:** This range is completely absorbed by the atmosphere and not employed in remote sensing.
3. **Ultraviolet**: **0.03–0.40 μm**
   i) **Hard UV 0.03–0.3 μm:** This range is completely absorbed by the atmosphere and not employed in remote sensing.
   ii) **Photographic UV 0.30–0.40 μm**: This range is not absorbed by the atmosphere and detectable with film and photo detectors but with severe atmospheric scattering.
4. **Visible Light**: This narrow band of electromagnetic radiation extends from about 400 nm (violet) to about 700 nm (red). It's available for remote sensing the Earth, can be imaged with photographic film.

Violet: 400-430 nm

Indigo: 430-450 nm

Blue: 450-500 nm: Because water increasingly absorbs electromagnetic (EM) radiation at longer wavelengths, band 1 provides the best data for mapping depth 15 detail of water-covered areas. It is also used for soil-vegetation discrimination, forest mapping, and distinguishing cultural features.

Green: 500-570 nm: The blue-green region of the spectrum corresponds to the chlorophyll absorption of healthy vegetation and is useful for mapping detail, such as depth or sediment in water bodies. Cultural features, such as roads and buildings also show up well in this band.

Yellow: 570-590 nm

Orange: 590-610 nm

Red: 610-700 nm: Chlorophyll absorbs these wavelengths in healthy vegetation. Hence, this band is useful for distinguishing plant species, as well as soil and geologic boundaries.

5. **Infrared**: 0.7 to 300 μm wavelength. This region is sensitive to plant water content, which is a useful measure in studies of vegetation health. This band is also used for distinguishing clouds, snow, and ice, mapping geologic formations and soil boundaries. It is also responsive to plant and soil moisture content. This region is further divided into the following bands:
   a) **Near Infrared (NIR)**: 0.7 to 1.5 μm.
   b) **Short Wavelength Infrared (SWIR)**: 1.5 to 3 μm.
   c) **Mid Wavelength Infrared (MWIR)**: 3 to 8 μm.
   d) **Long Wavelength Infrared (LWIR)**: 8 to 15 μm.
   e) **Far Infrared (FIR)**: longer than 15 μm.

The **NIR** and **SWIR** are also known as the ***Reflected Infrared***, referring to the main infrared component of the solar radiation reflected from the earth's surface. The **MWIR** and **LWIR** are the *Thermal Infrared*.

6. **Microwaves (Radar) 1 mm to 1 m wavelength**. Microwaves can penetrate clouds, fog, and rain. Images can be acquired in the active or passive mode. Radar is the active form of microwave remote sensing. Radar images are acquired at various wavelength bands.

7. **Radio and TV Waves**: 10 cm to 10 km wavelength. ***The longest-wavelength portion of the electromagnetic spectrum.***

Applications of Remote Sensing

There are probably hundreds of applications - these are typical:

- **Meteorology** - Study of atmospheric temperature, pressure, water vapour, etc..
- **Oceanography** - Measuring sea surface temperature, mapping ocean currents, and wave energy spectra and depth sounding of coastal and ocean depths.
- **Glaciology**- Measuring ice cap volumes, ice stream velocity, and sea ice distribution. **(Glacial)**
- **Geology**- Identification of rock type, mapping faults and structure.
- **Geodesy**- Measuring the figure of the Earth and its gravity field.
- **Topography and cartography** - Improving digital elevation models.
- **Agriculture-**Monitoring the biomass of land vegetation
- **Forest-** Monitoring the health of crops, mapping soil moisture
- **Botany**- Forecasting crop yields.
- **Hydrology-** Assessing water resources from snow, rainfall and underground aquifers.
- **Disaster warning and assessment -** Monitoring of floods and landslides, monitoring volcanic activity, assessing damage zones from natural disasters.
- **Planning applications -** Mapping ecological zones, monitoring deforestation, monitoring urban land use.
- **Oil and mineral exploration -** Locating natural oil seeps and slicks, mapping geological structures, monitoring oil field subsidence.
- **Military-** Developing precise maps for planning, monitoring military infrastructure, monitoring ship and troop movements
- **Urban**- Determining the status of a growing crop

- **Climate**- The effects of climate change on glaciers and Arctic and Antarctic regions
- **Sea**- Monitoring the extent of flooding
- **Rock-** Recognizing rock types
- **Space program**- This is the backbone of the space program
- **Seismology** - As a premonition.

Electromagnetic Radiation and the Electromagnetic Spectrum

EMR is a dynamic form of energy that propagates as wave motion at a velocity of $\mathbf{c = 3 \times 10^{10}}$ **cm/sec**. The parameters that characterize a wave motion are wavelength (λ), frequency (ν) and velocity (c) (Fig.). The relationship between the above is $\mathbf{c = \nu\lambda}$**.**

$$c = \nu\lambda.$$

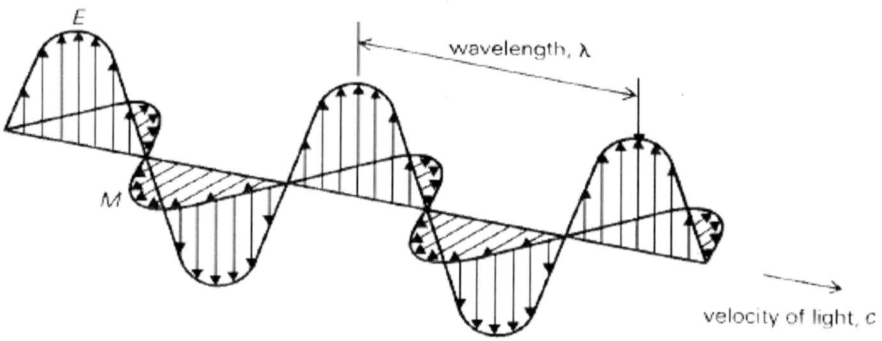

Fig. Electromagnetic wave. It has two components, Electric field E and Magnetic field M, both perpendicular to the direction of propagation

Electromagnetic energy radiates in accordance with the basic wave theory. This theory describes the EM energy as travelling in a harmonic sinusoidal fashion at the velocity of light. Although many characteristics of EM energy are easily described by wave theory, another theory known as particle theory offers insight into how electromagnetic energy interacts with matter. It suggests that EMR is composed of many discrete units called photons/quanta.

The energy of photon is $\mathbf{Q = hc/\lambda = h\nu}$, Where Q is the energy of quantum, h = Planck's constant.

Table 2: Principal Divisions of the Electromagnetic Spectrum

Wavelength	Description
Gamma rays	Gamma rays
X-rays	**X-rays**
Ultraviolet (UV) region 0.30 μm - 0.38 μm (1 μm = 10^6 m)	This region is beyond the violet portion of the visible wavelength, and hence Its name. Some earth's surface material primarily rocks and minerals emit visible UV radiation. However, UV radiation Is largely scattered by earths atmosphere and hence, not used In field of remote sensing.
Visible Spectrum 0.4 μm - 0.7 μm Violet 0.4 μm - 0.446 μm Blue 0.446 μm - 0.5 μm Green 0.5 μm - 0.578 μm Yellow 0.578 μm - 0.592 μm Orange 0.592 μm - 0.62 μm Red 0.62 μm -0.7 μm	This is the light, which our eyes can detect. This is the only portion of the spectrum that can be associated with the concept of colour. Blue Green and Red are the three primary colours of the visible spectrum. They are defined as such because no single primary colour can be created from the other two. but all other colours can be formed by combining the three in various proportions. The colour of an object is defined by the colour of the light It reflects.
Infrared (IR) Spectrum 0.7 μm - 100 μm	Wavelengths longer than the red portion of the visible spectrum are designated as the Infrared spectrum. British Astronomer/ VIIllam Herschel discovered this In 1800. The Infrared region can be divided Into two categories based on their radiation properties. Reflected IR (.7 μm - 3.0 μm) is used for remote sensing. Thermal IR (3 μm - 35 μm) Is the radiation emitted from earth's surface In the form of heat and used for remote sensing.
Microwave Region 1 mm - 1 m	This is the longest wavelength used in remote sensing. The shortest wavelengths In this range have properties similar to thermal Infrared region. The main advantage of this spectrum Is Its ability to penetrate through clouds.
Radio Waves (>1 m)	This is the longest portion of the spectrum mostly used for commercial broadcast and meteorology.

Remote sensing can be either passive or active. ACTIVE systems have their own source of energy (such as RADAR), whereas the PASSIVE systems depend upon external source of illumination (such as SUN) or self-emission for remote sensing.

Interaction of EMR with the Earth's Surface

Radiation from the sun, when incident upon the earth's surface, is either reflected by the surface, transmitted into the surface or absorbed and emitted by the surface (Fig. 3). The **EMR**, on interaction, experiences a number of changes in magnitude, direction, wavelength, polarization and phase. These changes are detected by the remote sensor and enable the interpreter to obtain useful information about the object of interest. The remotely sensed data contain both spatial information (size, shape and orientation) and spectral information (tone, colour and spectral signature).

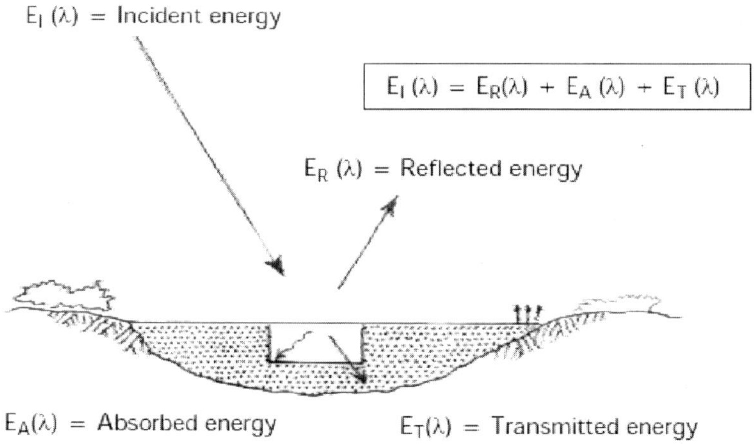

Fig. Interaction of Energy with the earth's surface. (*Source:* Liliesand & Kiefer, 1993)

From the viewpoint of interaction mechanisms, with the object-visible and infrared wavelengths from 0.3 μm to 16 μm can be divided into three regions. The spectral band from 0.3 μm to 3 μm is known as the reflective region. In this band, the radiation sensed by the sensor is that due to the sun, reflected by the earth's surface. The band corresponding to the atmospheric window between 8 μm and 14 μm is known as the thermal infrared band. The energy available in this band for remote sensing is due to thermal emission from the earth's surface. Both reflection and self-emission are important in the intermediate band from 3 μm to 5.5 μm. In the microwave region of the spectrum, the sensor is radar, which is an active sensor, as it provides its own source of **EMR**. The **EMR** produced by the radar is transmitted to the earth's surface and the **EMR** reflected (back scattered) from the surface is recorded and analyzed. The microwave region can also be monitored with passive sensors, called microwave radiometers, which record the radiation emitted by the terrain in the microwave region.

Reflection

Of all the interactions in the reflective region, surface reflections are the most useful and revealing in remote sensing applications. Reflection occurs when a ray of light is redirected, as it strikes a non-transparent surface. The reflection intensity depends on the surface refractive index, absorption coefficient and the angles of incidence and reflection (Fig. 4).

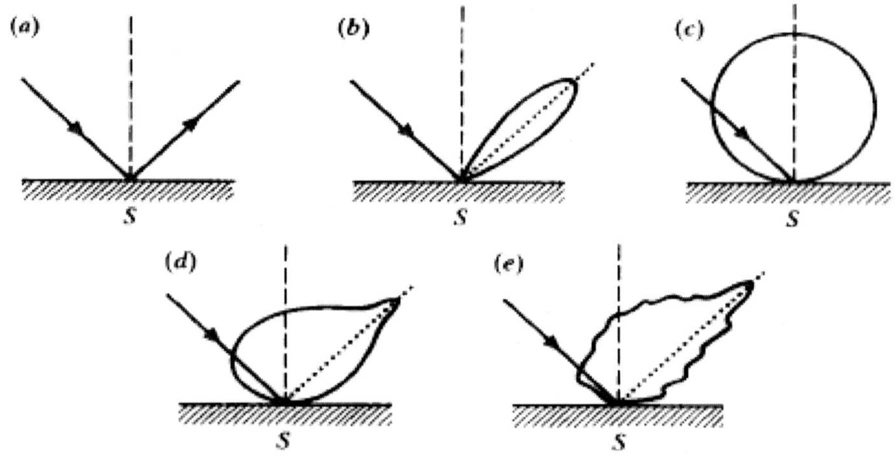

Fig. Different types of scattering surfaces (a) Perfect specular reflector (b) Near perfect specular reflector (c) Lambertain (d) Quasi-Lambertian (e) Complex.

Transmission

Transmission of radiation occurs when radiation passes through a substance without significant attenuation. For a given thickness, or depth of a substance, the ability of a medium to transmit energy is measured as transmittance (τ).

$$\tau = \frac{\text{Transmitted radiation}}{\text{Incident radiation}}$$

Spectral Signature

Spectral reflectance, [$\rho(\lambda)$], is the ratio of reflected energy to incident energy as a function of wavelength. Various materials of the earth's surface have different spectral reflectance characteristics. Spectral reflectance is responsible for the colour or tone in a photographic image of an object. Trees appear green because they reflect more of the green wavelength. The values of the spectral reflectance of objects averaged over different, well-defined wavelength intervals comprise the spectral signature of the objects or features by which they can be distinguished. To obtain the necessary ground truth for the interpretation of multispectral imagery, the spectral characteristics of various natural objects have been extensively measured and recorded.

The spectral reflectance is dependent on wavelength, it has different values at different wavelengths for a given terrain feature. The reflectance characteristics of the earth's surface features are expressed by spectral reflectance, which is given by:

$$\rho(\lambda) = \left[ER(\lambda)/EI(\lambda)\right] \times 100$$

Where,

$\rho(\lambda)$ = Spectral reflectance (reflectivity) at a particular wavelength.

ER (λ) = Energy of wavelength reflected from object

EI (λ) = Energy of wavelength incident upon the object

The plot between $\rho(\lambda)$ and (λ) is called a spectral reflectance curve. This varies with the variation in the chemical composition and physical conditions of the feature, which results in a range of values. The spectral response patterns are averaged to get a generalized form, which is called generalized spectral response pattern for the object concerned. Spectral signature is a term used for unique spectral response pattern, which is characteristic of a terrain feature. Figure 5 shows a typical reflectance curves for three basic types of earth surface features, healthy vegetation, dry bare soil (grey-brown and loamy) and clear lake water.

Reflectance Characteristics of Earth's Cover types

The spectral characteristics of the three main earth surface features are discussed below :

Vegetation : The spectral characteristics of vegetation vary with wavelength. Plant pigment in leaves called chlorophyll strongly absorbs radiation in the red and blue wavelengths but reflects green wavelength. The internal structure of healthy leaves acts as diffuse reflector of near infrared wavelengths. Measuring and monitoring the near infrared reflectance is one way that scientists determine how healthy particular vegetation may be.

Water: Majority of the radiation incident upon water is not reflected but is either absorbed or transmitted. Longer visible wavelengths and near infrared radiation is absorbed more by water than by the visible wavelengths. Thus, water looks blue or blue green due to stronger reflectance at these shorter wavelengths and darker, if viewed at red or near infrared wavelengths. The factors that affect the variability in reflectance of a water body are depth of water, materials within water and surface roughness of water.

Soil: The majority of radiation incident on a soil surface is either reflected or absorbed and little is transmitted. The characteristics of soil that determine its reflectance properties are its moisture content, organic matter content, texture, structure and iron oxide content. The soil curve shows less peak and valley variations. The presence of moisture in soil decreases its reflectance. By measuring the energy that is reflected by targets on earth's surface over a variety of different wavelengths, we can build up a spectral signature for that object. And by comparing the response pattern of different features, we may be able to distinguish between them, which we may not be able to do if we only compare them at one wavelength. For example, Water and Vegetation reflect somewhat similarly in the visible wavelength but not in the infrared.

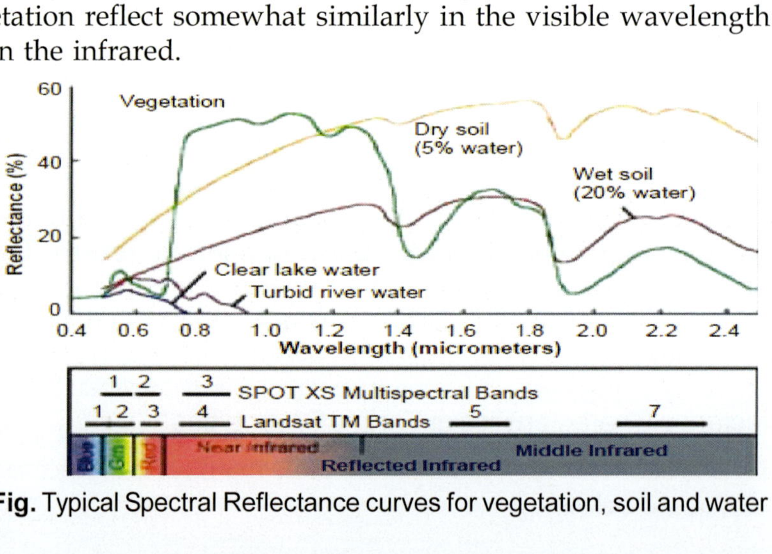

Fig. Typical Spectral Reflectance curves for vegetation, soil and water

Interactions with the Atmosphere

The sun is the source of radiation, and electromagnetic radiation (EMR) from the sun that is reflected by the earth and detected by the satellite or aircraft-borne sensor must pass through the atmosphere twice, once on its journey from the sun to the earth and second after being reflected by the surface of the earth back to the sensor. Interactions of the direct solar radiation and reflected radiation from the target with the atmospheric constituents interfere with the process of remote sensing and are called as **"Atmospheric Effects".**

The interaction of EMR with the atmosphere is important to remote sensing for two main reasons. First, information carried by EMR reflected/emitted by the earth's surface is modified while traversing through the atmosphere. Second, the interaction of EMR with the atmosphere can be used to obtain useful information about the atmosphere itself. The atmospheric constituents scatter and absorb the radiation modulating the radiation reflected from the target by attenuating it, changing its spatial distribution and introducing into field of view radiation from sunlight scattered in the atmosphere and some of the energy reflected from nearby ground area. Both scattering and absorption vary in their effect from one part of the spectrum to the other.

The solar energy is subjected to modification by several physical processes as it passes the atmosphere, viz.

1. Scattering;
2. Absorption; and
3. Refraction

1. Scattering

Atmospheric Scattering consists of three types of scattering:

a) Rayleigh Scattering

b) Mie Scattering

c) Non-selective Scattering

Atmospheric Scattering

Scattering is the redirection of EMR by particles suspended in the atmosphere or by large molecules of atmospheric gases. Scattering not only reduces the image contrast but also changes the spectral signature of ground objects as seen by the sensor. The amount of scattering depends upon the size of the particles, their abundance, the wavelength of radiation, depth of the atmosphere through which the energy is travelling and the concentration of the particles. The concentration of particulate matter varies both in time and over season. Thus, the effects of scattering will be uneven spatially and will vary from time to time.

Theoretically scattering can be divided into three categories depending upon the wavelength of radiation being scattered and the size of the particles causing the scattering. The three different types of scattering from particles of different sizes are summarized below:

Scattering process	Wavelength	Approximate dependence particle size	Kinds of particles
Selective			
• Rayleigh	λ^{-4}	< 1 μm	Air molecules
• Mie	λ^{o} to λ^{-4}	0.1 to 10 μm	Smoke, haze
• Non-selective	λ^{o}	> 10 μm	Dust, fog, clouds

a) Rayleigh Scattering

Rayleigh scattering predominates where electromagnetic radiation interacts with particles that are smaller than the wavelength of the incoming light. The effect of the Rayleigh scattering is inversely proportional to the fourth power of the wavelength. Shorter wavelengths are scattered more than longer wavelengths. In the absence of these particles and scattering the sky would appear black. In the context of remote sensing, the Rayleigh scattering is the most important type of scattering. It causes a distortion of spectral characteristics of the reflected light, when compared to measurements taken on the ground.

b) Mie Scattering

Mie scattering occurs, when the wavelength of the incoming radiation is similar in size to the atmospheric particles. These are caused by aerosols: a mixture of gases, water vapour and dust. It is generally restricted to the lower atmosphere where the larger particles are abundant and dominates under overcast cloud conditions. It influences the entire spectral region from ultraviolet to near infrared regions.

c) Non-selective Scattering

This type of scattering occurs, when the particle size is much larger than the wavelength of the incoming radiation. Particles responsible for this effect are water droplets and larger dust particles. The scattering is independent of the wavelength, all the wavelength are scattered equally. The most common example of non-selective scattering is the appearance of clouds as white. As clouds consist of water droplet particles and the wavelengths are scattered in equal amount, the clouds appears as white.

Occurrence of this scattering mechanism gives a clue to the existence of large particulate matter in the atmosphere above the scene of interest which itself is a useful data. Using minus blue filters can eliminate the effects of the Rayleigh component of scattering. However, the effect of heavy haze i.e. when all the wavelengths are scattered uniformly, cannot be eliminated using haze filters. The effects of haze are less pronounced in the thermal infrared region. Microwave radiation is completely immune to haze and can even penetrate clouds.

2. Atmospheric Absorption

The gas molecules present in the atmosphere strongly absorb the EMR passing through the atmosphere in certain spectral bands. Mainly three gases are responsible for most of absorption of solar radiation, viz. ozone, carbon dioxide and water vapour. Ozone absorbs the high energy, short wavelength portions of the ultraviolet spectrum ($\lambda < 0.24$ μm) thereby preventing the transmission of this radiation to the lower atmosphere. Carbon dioxide is important in remote sensing, as it effectively absorbs the radiation in mid and far infrared regions of the spectrum. It strongly absorbs in the region from about 13-17.5 μm, whereas two most important regions of water vapour absorption are in bands 5.5 - 7.0 μm and above 27 μm. Absorption relatively reduces the amount of light that reaches our eye making the scene look relatively duller.

Atmospheric Windows

The general atmospheric transmittance across the whole spectrum of wavelengths is shown in Figure 6. The atmosphere selectively transmits energy of certain wavelengths. **The spectral bands for which the atmosphere is relatively transparent are known as atmospheric windows.** Atmospheric windows are present in the visible part (0.4 μm - 0.76 μm) and the infrared regions of the EM spectrum. In the visible part transmission is mainly effected by ozone absorption and by molecular scattering. The atmosphere is transparent again beyond about λ= 1mm, the region used for microwave remote sensing.

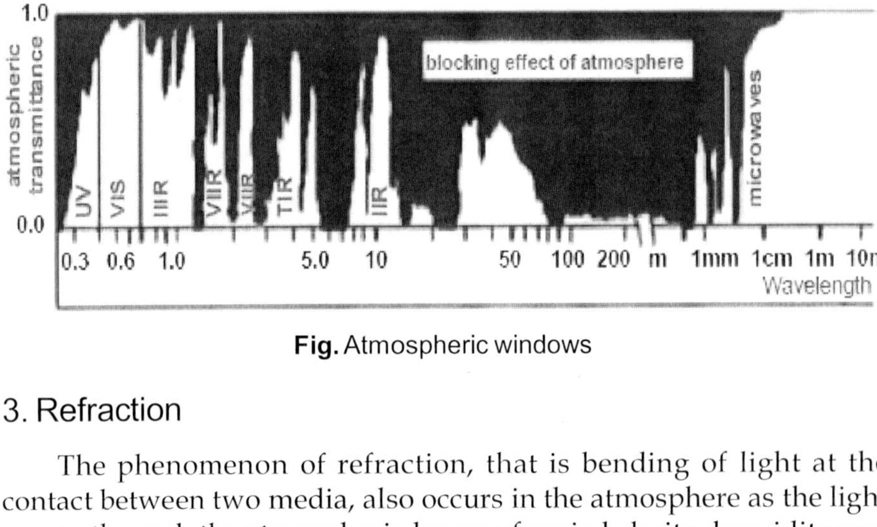

Fig. Atmospheric windows

3. Refraction

The phenomenon of refraction, that is bending of light at the contact between two media, also occurs in the atmosphere as the light passes through the atmospheric layers of varied clarity, humidity and temperature. These variations influence the density of atmospheric layers, which in turn, causes the bending of light rays as they pass from one layer to another. The most common phenomena are the mirage like apparitions sometimes visible in the distance on hot summer days.

Conclusions

Remote sensing technology has developed from balloon photography to aerial photography to multi-spectral satellite imaging. Radiation interaction characteristics of the earth and atmosphere in different regions of electromagnetic spectrum are very useful for identifying and characterizing the earth and atmospheric features.

Electromagnetic Radiation

As was noted in the previous section, the first requirement for remote sensing is to have an **energy source to illuminate the target** (unless the sensed energy is being emitted by the target). This energy is in the form of electromagnetic radiation.

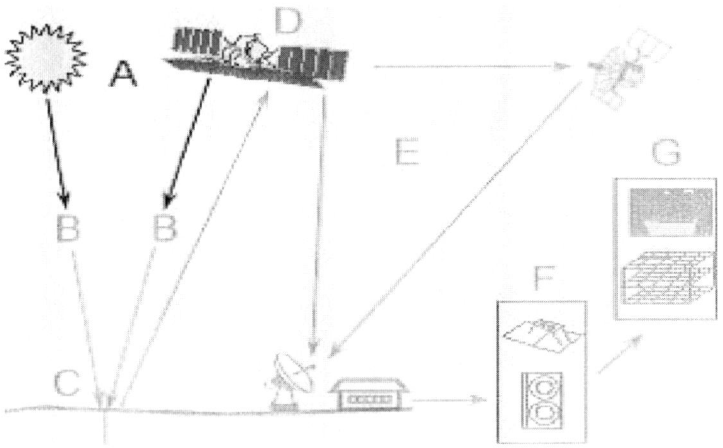

All electromagnetic radiation has fundamental properties and behaves in predictable ways according to the basics of wave theory.

Electromagnetic radiation consists of an **electrical field (E)** which varies in magnitude in a direction perpendicular to the direction in which the radiation is travelling, and a **magnetic field (M)** oriented at right angles to the electrical field. Both these fields travel at the speed of light (c).

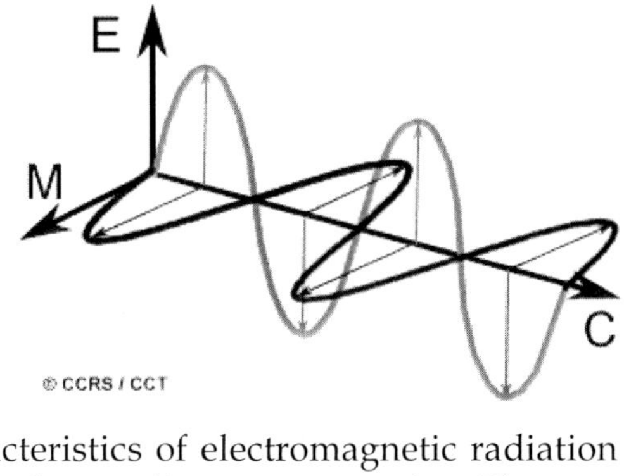

Two characteristics of electromagnetic radiation are particularly important for understanding remote sensing. These are the **wavelength and frequency.**

The wavelength is the length of one wave cycle, which can be measured as the distance between successive wave crests. Wavelength is usually represented by the Greek letter lambda (ë). Wavelength is measured in metres (m) or some factor of metres, such as **nanometres** (nm, 10-9 metres), **micrometres** (μm, 10-6 metres) (μm, 10-6 metres) or centimetres (cm, 10-2 metres). Frequency refers to the number of cycles of a wave passing a fixed point per unit of time. Frequency is normally measured in **hertz** (Hz), equivalent to one cycle per second, and various multiples of hertz.

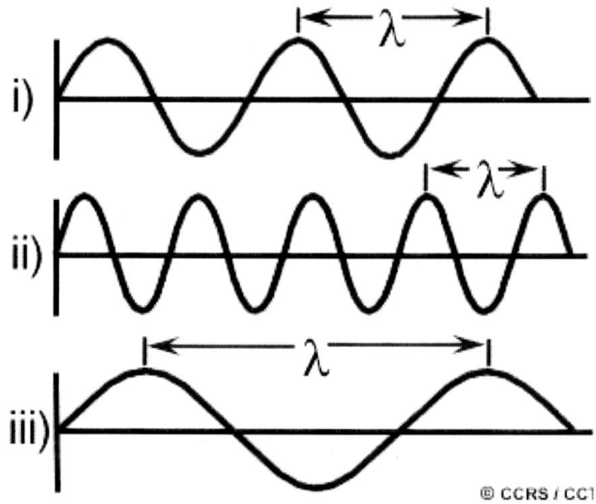

Wavelength and frequency are related by the following formula:

$c = \lambda v$

Where:

λ = wavelength(m)

v = frequency (cycles per second, Hz)

c= speed of light (3 x 10^8m/s)

Therefore, the two are inversely related to each other. The shorter the wavelength, the higher the frequency. The longer the wavelength, the lower the frequency. Understanding the characteristics of electromagnetic radiation in terms of their wavelength and frequency is crucial to understanding the information to be extracted from remote sensing data. Next we will be examining the way in which we categorize electromagnetic radiation for just that purpose.

The Electromagnetic Spectrum

The **electromagnetic spectrum** ranges from the shorter wavelengths (including gamma and x-rays) to the longer wavelengths (including microwaves and broadcast radio waves). There are several regions of the electromagnetic spectrum which are useful for remote sensing. For most purposes, the **ultraviolet or UV** portion of the spectrum has the shortest wavelengths which are practical for remote sensing. This radiation is just beyond the violet portion of the visible wavelengths, hence its name. Some Earth surface materials, primarily rocks and minerals, fluoresce or emit visible light, when illuminated by UV radiation.

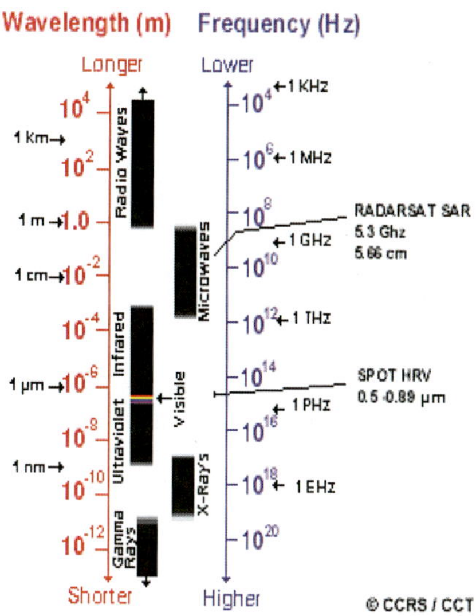

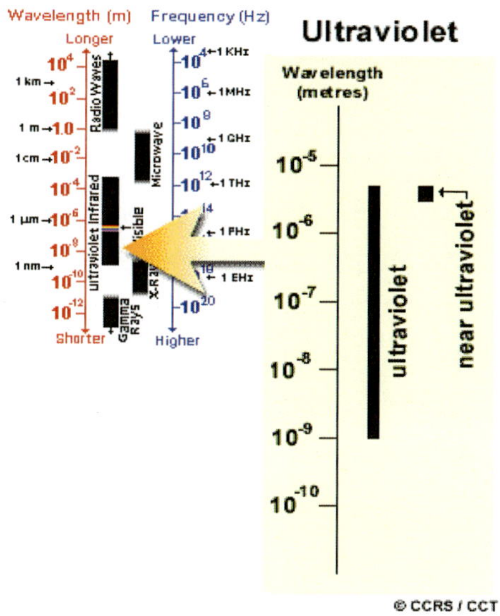

The light which our eyes - our "remote sensors" - can detect is part of the **visible spectrum**. It is important to recognize how small the visible portion is relative to the rest of the spectrum. There is a lot of radiation around us which is "invisible" to our eyes, but can be detected by other remote sensing instruments and used to our advantage. The visible wavelengths cover a range from approximately 0.4 to 0.7 µm. The longest visible wavelength is red and the shortest is violet. Common wavelengths of what we perceive as particular colours from the visible portion of the spectrum are listed below. It is important to note that this is the only portion of the spectrum, we can associate with the concept of **colours**.

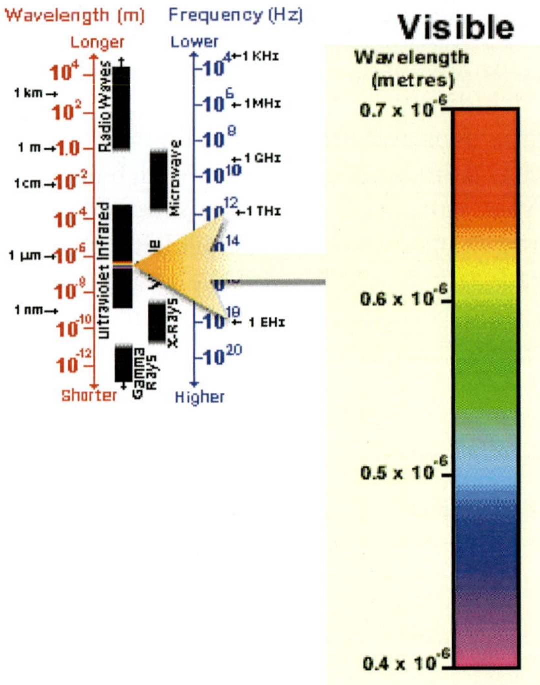

- **Violet:** 0.4 - 0.446 μm
- **Blue:** 0.446 - 0.500 μm
- **Green:** 0.500 - 0.578 μm
- **Yellow:** 0.578 - 0.592 μm
- **Orange:** 0.592 - 0.620 μm
- **Red:** 0.620 - 0.7 μm

Blue, **green**, and **red** are the **primary colours** or wavelengths of the visible spectrum. They are defined as such because no single primary colour can be created from the other two, but all other colours can be formed by combining blue, green, and red in various proportions. Although we see sunlight as a uniform or homogeneous colour, it is actually composed of various wavelengths of radiation in primarily the ultraviolet, visible and infrared portions of the spectrum. The visible portion of this radiation can be shown in its component colours when sunlight is passed through a **prism**, which bends the light in differing amounts according to wavelength.

The next portion of the spectrum of interest is the infrared (IR) region which covers the wavelength range from approximately 0.7 μm to 100 μm - more than 100 times as wide as the visible portion! The infrared region can be divided into two categories based on their radiation properties- the **reflected IR**, and the emitted or **thermal IR**. Radiation in the reflected IR region is used for remote sensing purposes in ways very similar to radiation in the visible portion. The reflected IR covers wavelengths from approximately 0.7 μm to 3.0 μm. The thermal IR region is quite different than the visible and reflected IR portions, as this energy is essentially the radiation that is emitted from the Earth's surface in the form of heat. The thermal IR covers wavelengths from approximately 3.0 μm to 100 μm.

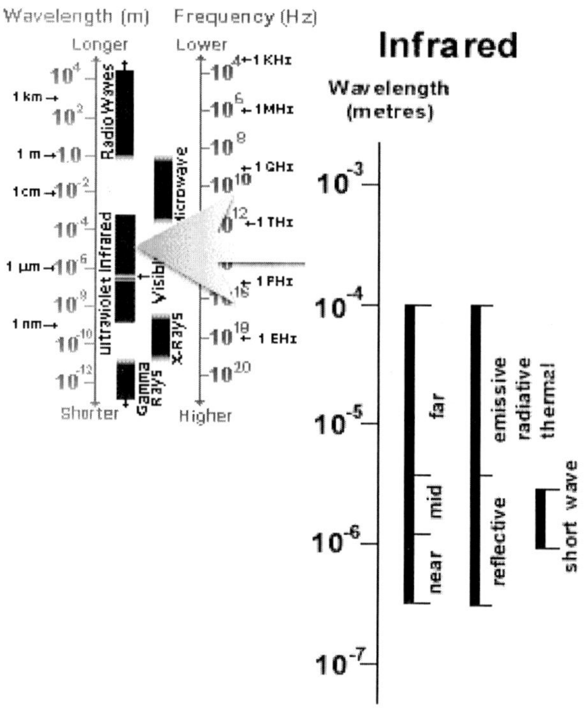

The portion of the spectrum of more recent interest to remote sensing is the **microwave region** from about 1 mm to 1 m. This covers the longest wavelengths used for remote sensing. The shorter wavelengths have properties similar to the thermal infrared region while the longer wavelengths approach the wavelengths used for radio broadcasts. Because of the special nature of this region and its importance to remote sensing in Canada, an entire chapter (Chapter 3) of the tutorial is dedicated to microwave sensing.

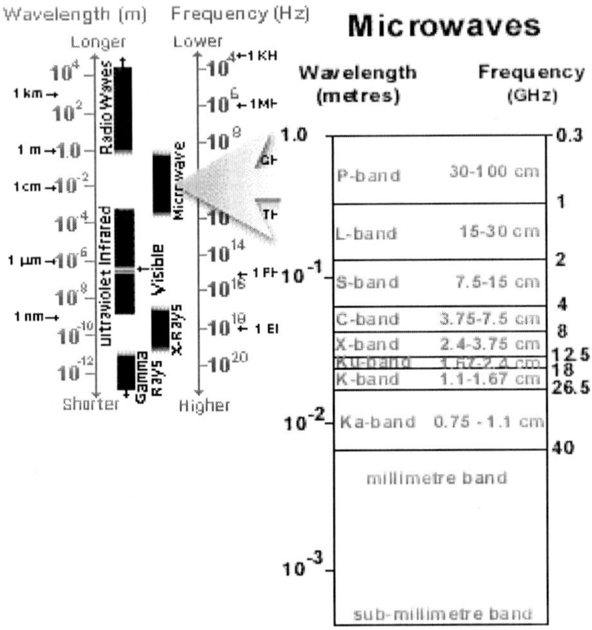

Interactions with the Atmosphere

Before radiation used for remote sensing reaches the Earth's surface, it has to travel through some distance of the Earth's atmosphere. Particles and gases in the atmosphere can affect the incoming light and radiation. These effects are caused by the mechanisms of **scattering** & **absorption**.

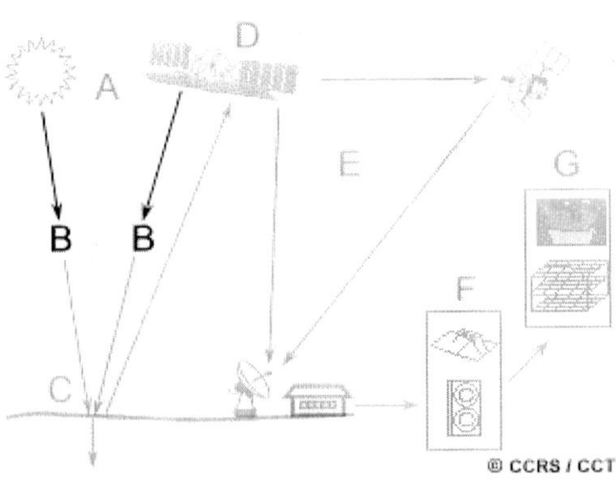

Scattering occurs, when particles or large gas molecules present in the atmosphere interact with and cause the electromagnetic radiation to be redirected from its original path. How much scattering takes place depends on several factors including the wavelength of the radiation, the abundance of particles or gases, and the distance the radiation travels through the atmosphere. There are three (3) types of scattering which take place.

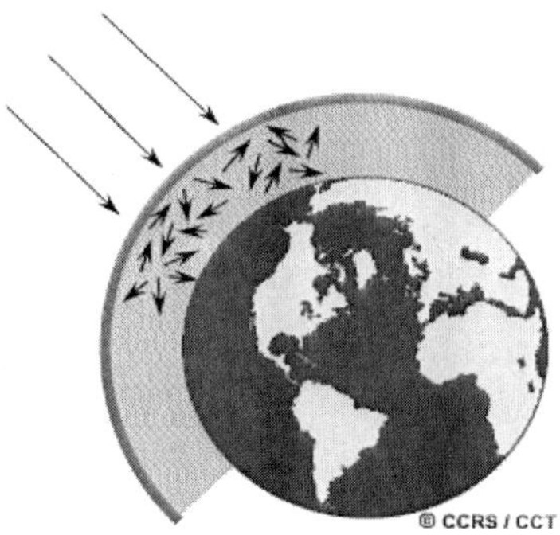

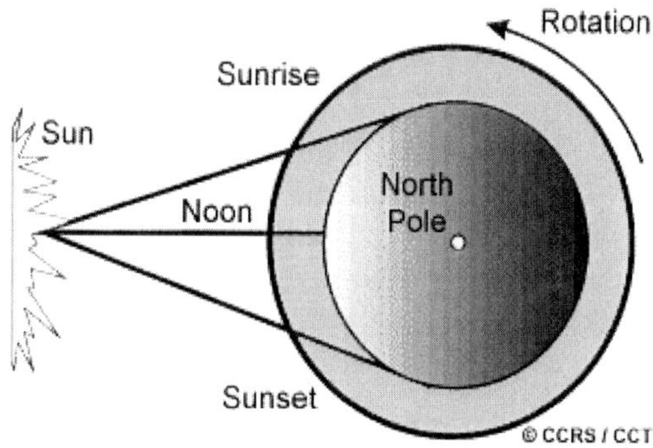

Rayleigh scattering occurs, when particles are very small compared to the wavelength of the radiation. These could be particles, such as small specks of dust or nitrogen and oxygen molecules. Rayleigh scattering causes shorter wavelengths of energy to be scattered much more than longer wavelengths. Rayleigh scattering is the dominant scattering mechanism in the upper atmosphere. The fact that the sky appears "blue" during the day is because of this phenomenon. As sunlight passes through the atmosphere, the shorter wavelengths (i.e. blue) of the visible spectrum are scattered more than the other (longer) visible wavelengths. At **sunrise and sunset,** the light has to travel farther through the atmosphere than at midday and the scattering of the shorter wavelengths is more complete; this leaves a greater proportion of the longer wavelengths to penetrate the atmosphere.

Mie scattering occurs, when the particles are just about the same size as the wavelength of the radiation. Dust, pollen, smoke and water vapour are common causes of Mie scattering which tends to affect longer wavelengths than those affected by Rayleigh scattering. Mie scattering occurs mostly in the lower portions of the atmosphere where larger particles are more abundant, and dominates, when cloud conditions are overcast.

The final scattering mechanism of importance is called **non-selective scattering**. This occurs, when the particles are much larger than the wavelength of the radiation. Water droplets and large dust particles can cause this type of scattering. Non-selective scattering gets its name from the fact that all wavelengths are scattered about equally. This type of scattering causes fog and clouds to appear white to our eyes because blue, green, and red light are all scattered in approximately equal quantities (blue+green+red light = white light).

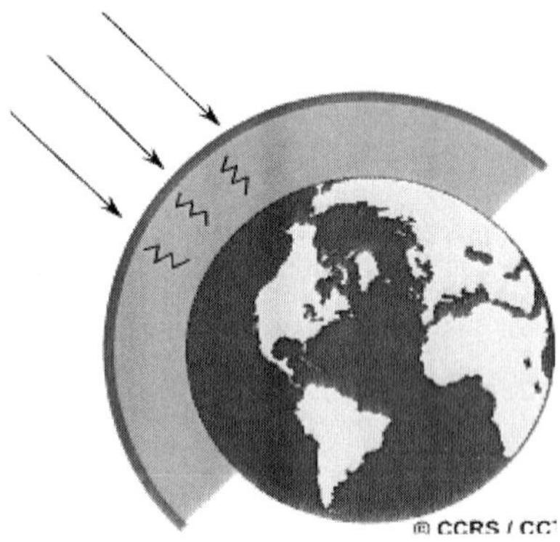

Absorption is the other main mechanism at work when electromagnetic radiation interacts with the atmosphere. In contrast to scattering, this phenomenon causes molecules in the atmosphere to absorb energy at various wavelengths. Ozone, carbon dioxide, and water vapour are the three main atmospheric constituents which absorb radiation. **Ozone** serves to absorb the harmful (to most living things) ultraviolet radiation from the sun. Without this protective layer in the atmosphere our skin would burn, when exposed to sunlight.

You may have heard **carbon dioxide** referred to as a greenhouse gas. This is because it tends to absorb radiation strongly in the far infrared portion of the spectrum - that area associated with thermal heating - which serves to trap this heat inside the atmosphere. Water vapour in the atmosphere absorbs much of the incoming longwave infrared and shortwave microwave radiation (between 22μm and 1m). The presence of water vapour in the lower atmosphere varies greatly from location to location and at different times of the year. For example, the air mass above a desert would have very little water vapour to absorb energy, while the tropics would have high concentrations of water vapour (i.e. high humidity).

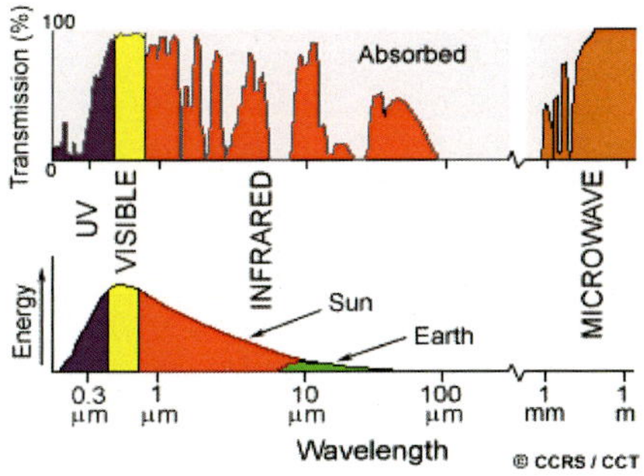

Because these gases absorb electromagnetic energy in very specific regions of the spectrum, they influence where (in the spectrum) we can "look" for remote sensing purposes. Those areas of the spectrum which are not severely influenced by atmospheric absorption and thus, are useful to remote sensors are called **atmospheric windows**.

By comparing the characteristics of the two most common energy/ radiation sources (the sun and the earth) with the atmospheric windows available to us, we can define those wavelengths that we can use **most effectively** for remote sensing. The visible portion of the spectrum, to which our eyes are most sensitive, corresponds to both an atmospheric window and the peak energy level of the sun. Note also that heat energy emitted by the Earth corresponds to a window around 10 μm in the thermal IR portion of the spectrum, while the large window at wavelengths beyond 1 mm is associated with the microwave region.

Radiation - Target Interactions

Radiation that is not absorbed or scattered in the atmosphere can reach and interact with the Earth's surface. There are three (3) forms of interaction that can take place when energy strikes, or is **incident (I)** upon the surface. These are: **absorption (A)**; **transmission (T)**; and **reflection (R)**.

The total incident energy will interact with the surface in one or more of these three ways. The proportions of each will depend on the wavelength of the energy and the material and condition of the feature.

Absorption (A) occurs, when radiation (energy) is absorbed into the target while transmission (T) occurs, when radiation passes through a target. Reflection (R) occurs, when radiation "bounces" off the target and is redirected. In remote sensing, we are most interested in measuring the radiation reflected from targets. We refer to two types of reflection, which represent the two extreme ends of the way in which energy is reflected from a target: **specular reflection** and **diffuse reflection**.

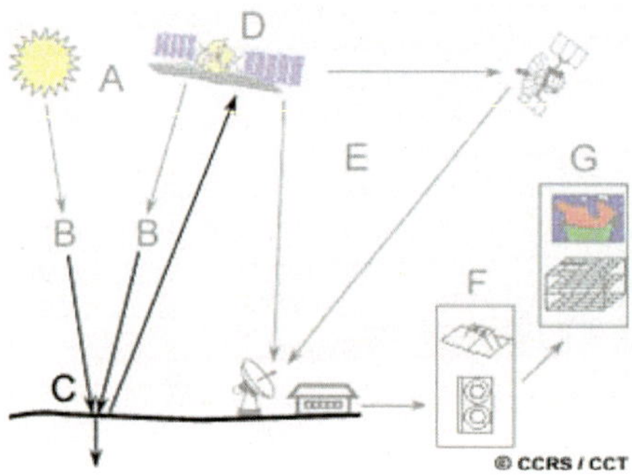

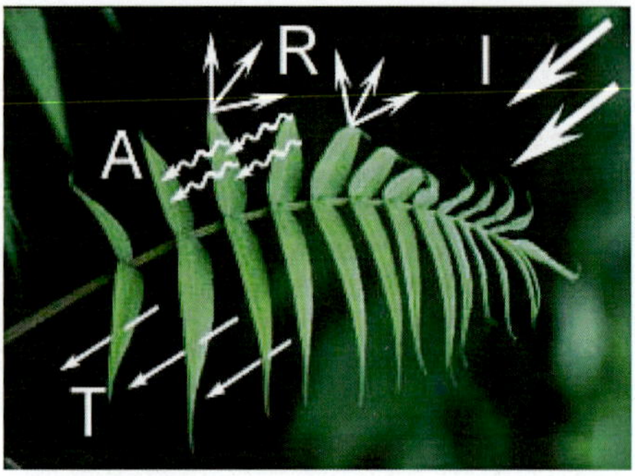

When a surface is smooth we get **specular** or mirror-like reflection where all (or almost all) of the energy is directed away from the surface in a single direction. **Diffuse** reflection occurs, when the surface is rough and the energy is reflected almost uniformly in all directions.

Most earth surface features lie somewhere between perfectly specular or perfectly diffuse reflectors. Whether a particular target reflects specularly or diffusely, or somewhere in between, depends on the surface roughness of the feature in comparison to the wavelength of the incoming radiation. If the wavelengths are much smaller than the surface variations or the particle sizes that make up the surface, diffuse reflection will dominate. For example, fine grained sand would appear fairly smooth to long wavelength microwaves but would appear quite rough to the visible wavelengths.

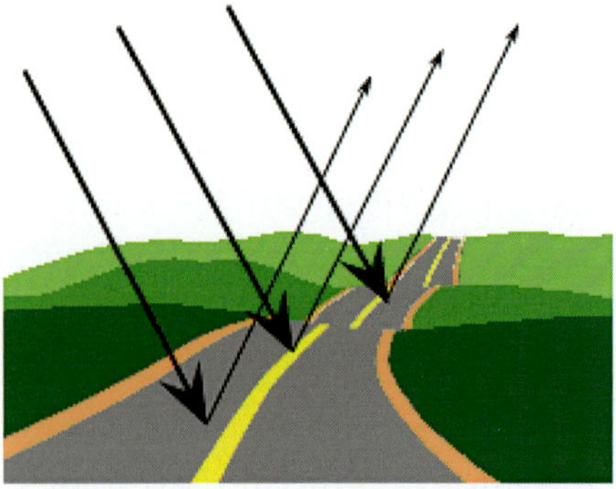

Let's take a look at a couple of examples of targets at the Earth's surface and how energy at the visible and infrared wavelengths interacts with them.

Leaves : A chemical compound in leaves called chlorophyll strongly absorbs radiation in the red and blue wavelengths but reflects green wavelengths. Leaves appear "greenest" to us in the summer, when chlorophyll content is at its maximum. In autumn, there is less chlorophyll in the leaves, so there is less absorption and proportionately more reflection of the red wavelengths, making the leaves appear red or yellow (yellow is a combination of red and green wavelengths).

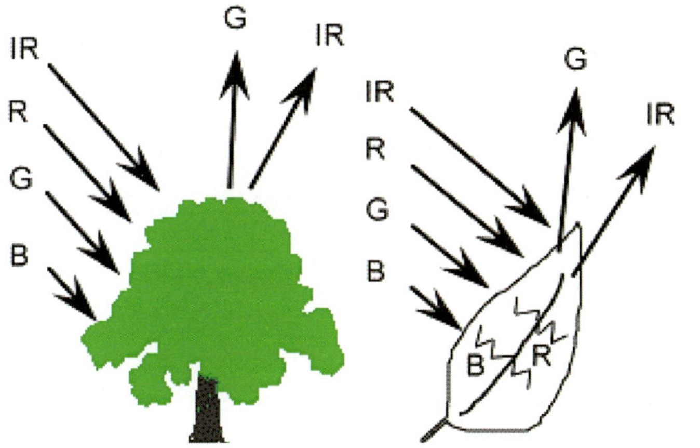

The internal structure of healthy leaves act as excellent diffuse reflectors of near-infrared wavelengths. If our eyes were sensitive to near-infrared, trees would appear extremely bright to us at these wavelengths. In fact, measuring and monitoring the near-IR reflectance is one way that scientists can determine how healthy (or unhealthy) vegetation may be.

Water: Longer wavelength visible and near infrared radiation is absorbed more by water than shorter visible wavelengths. Thus, water typically looks blue or blue-green due to stronger reflectance at these shorter wavelengths, and darker if viewed at red or near infrared wavelengths. If there is suspended sediment present in the upper layers of the water body, then this will allow better reflectivity and a brighter appearance of the water.

The apparent colour of the water will show a slight shift to longer wavelengths. Suspended sediment (S) can be easily confused with shallow (but clear) water, since these two phenomena appear very similar. Chlorophyll in algae absorbs more of the blue wavelengths and reflects the green, making the water appear more green in colour, when algae is present. The topography of the water surface (rough, smooth, floating materials, etc.) can also lead to complications for water-related interpretation due to potential problems of specular reflection and other influences on colour and brightness.

We can see from these examples that, depending on the complex make-up of the target that is being looked at, and the wavelengths of radiation involved, we can observe very different responses to the mechanisms of absorption, transmission, and reflection. By measuring the energy that is reflected (or emitted) by targets on the Earth's surface over a variety of different wavelengths, we can build up a **spectral response** for that object. By comparing the response patterns of different features, we may be able to distinguish between them, where we might not be able to, if we only compared them at one wavelength.

For example, water and vegetation may reflect somewhat similarly in the visible wavelengths but are almost always separable in the infrared. Spectral response can be quite variable, even for the same target type, and can also vary with time (e.g. "green-ness" of leaves) and location. Knowing where to "look" spectrally and understanding the factors which influence the spectral response of the features of interest are critical to correctly interpreting the interaction of electromagnetic radiation with the surface.

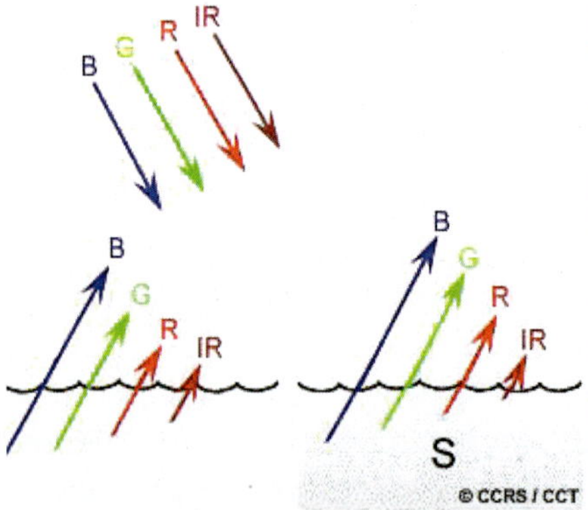

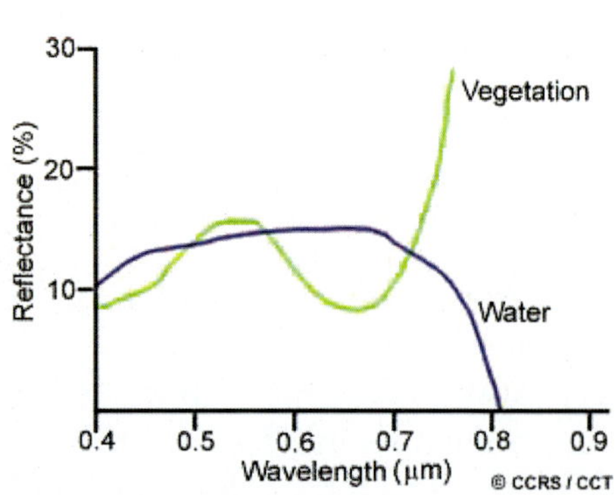

Passive Sensing vs. Active Sensing

So far, throughout this chapter, we have made various references to the sun as a source of energy or radiation. The sun provides a very convenient source of energy for remote sensing. The sun's energy is either **reflected**, as it is for visible wavelengths, or absorbed and then **reemitted**, as it is for thermal infrared wavelengths. Remote sensing systems which measure energy that is naturally available are called **passive sensors**. Passive sensors can only be used to detect energy, when the naturally occurring energy is available. For all reflected energy, this can only take place during the time when the sun is illuminating the Earth. There is no reflected energy available from the sun at night. Energy that is naturally emitted (such as thermal infrared) can be detected day or night, as long as the amount of energy is large enough to be recorded.

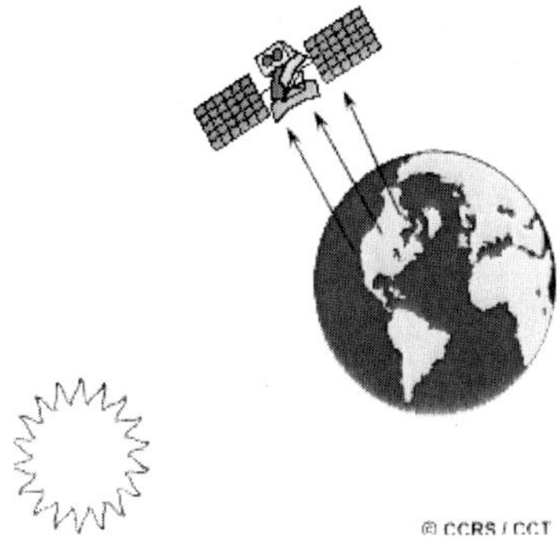

Active sensors, on the other hand, provide their own energy source for illumination. The sensor emits radiation which is directed toward the target to be investigated. The radiation reflected from that target is detected and measured by the sensor. Advantages for active sensors include the ability to obtain measurements anytime, regardless of the time of day or season.

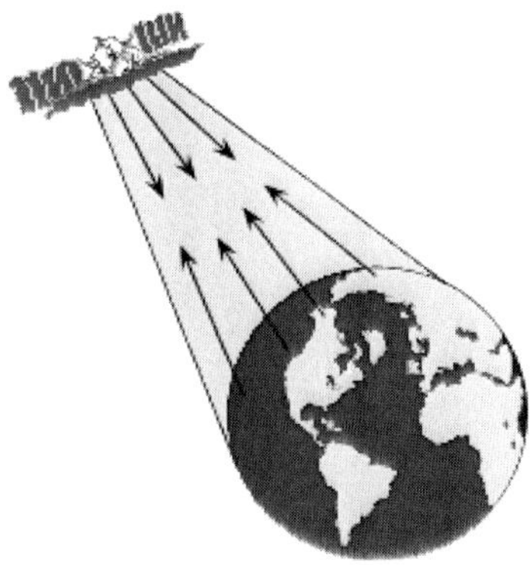

Active sensors can be used for examining wavelengths that are not sufficiently provided by the sun, such as microwaves, or to better control the way a target is illuminated. However, active systems require the generation of a fairly large amount of energy to adequately illuminate targets. Some examples of active sensors are a laser fluorosensor and a synthetic aperture radar (SAR).

Characteristics of Images

Electromagnetic energy may be detected either photographically or electronically. The photographic process uses chemical reactions on the surface of light-sensitive film to detect and record energy variations. It is important to distinguish between the terms **images** and **photographs** in remote sensing. An image refers to any pictorial representation, regardless of what wavelengths or remote sensing device has been used to detect and record the electromagnetic energy.

A **photograph** refers specifically to images that have been detected as well as recorded on photographic film. The black and white photo to the left, of part of the city of Ottawa, Canada was taken in the visible part of the spectrum. Photos are normally recorded over the wavelength range from 0.3 μm to 0.9 μm - the visible and reflected infrared. Based on these definitions, we can say that all photographs are images, but not all images are photographs. Therefore, unless we are talking specifically about an image recorded photographically, we use the term image.

A photograph could also be represented and displayed in a **digital** format by subdividing the image into small equal-sized and shaped areas, called picture elements or **pixels**, and representing the brightness of each area with a numeric value or **digital number**. Indeed, that is exactly what has been done to the photo to the left. In fact, using the definitions we have just discussed, this is actually a **digital image** of the original photograph!

The photograph was scanned and subdivided into pixels with each pixel assigned a digital number representing its relative brightness. The computer displays each digital value as different brightness levels. Sensors that record electromagnetic energy, electronically record the energy as an array of numbers in digital format right from the start.

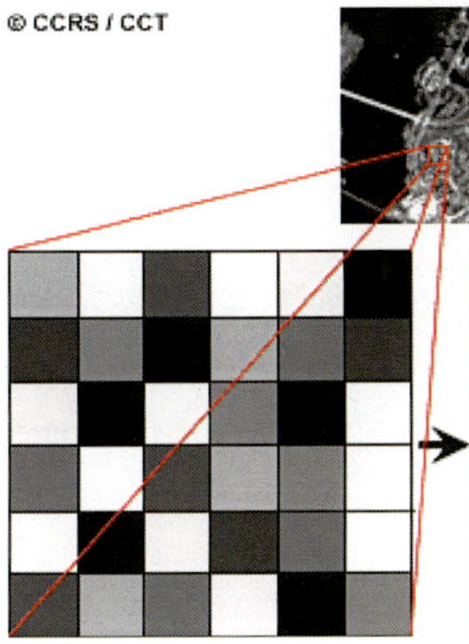

170	238	85	255	221	0
68	136	17	170	119	68
221	0	238	136	0	255
119	255	85	170	136	238
238	17	221	68	119	255
85	170	119	221	17	136

These two different ways of representing and displaying remote sensing data, either pictorially or digitally, are interchangeable, as they convey the same information (although some detail may be lost when converting back and forth). In previous sections, we described the visible portion of the spectrum and the concept of colours. We see colour because our eyes detect the entire visible range of wavelengths and our brains process the information into separate colours. Can you imagine what the world would look like, if we could only see very narrow ranges of wavelengths or colours? That is how many sensors work.

The information from a narrow wavelength range is gathered and stored in a **channel**, also sometimes referred to as a **band**. We can combine and display channels of information digitally using the three primary colours (blue, green, and red). The data from each channel is represented as one of the primary colours and, depending on the relative brightness (i.e. the digital value) of each pixel in each channel, the primary colours combine in different proportions to represent different colours. When we use this method to display a single channel or range of wavelengths, we are actually displaying that channel through all three primary colours. Because the brightness level of each pixel is the same for each primary colour, they combine to form a **black and white image**, showing various shades of grey from black to white. When we display more than one channel, each as a different primary colour, then the brightness levels may be different for each channel/primary colour combination and they will combine to form a **colour image**.

Some characteristics of Black body radiation (Four Laws of Light)

Black body : An ideal substance that absorbs all the radiant energy incident on it and emits radiant energy at the maximum possible rate per unit area at each wavelength for any given temperature. No actual substance behaves as a true black body.

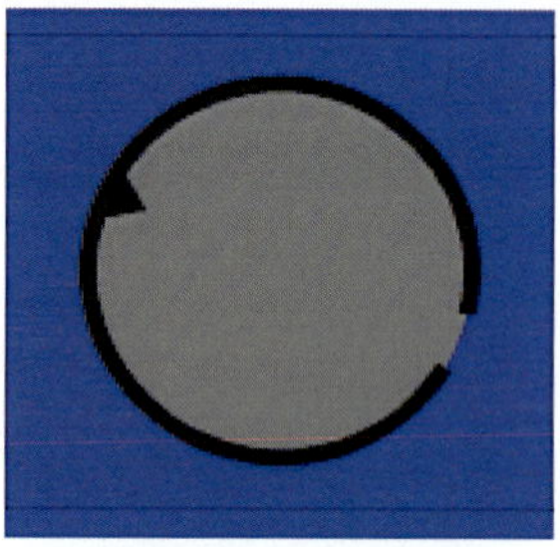

A black body is an idealized physical body that absorbs all incident electromagnetic radiation, regardless of frequency or angle of incidence. **A white body** is one with a "rough surface [that] reflects all incident rays completely and uniformly in all directions.

Black body radiation: The electromagnetic radiation emitted by an ideal blackbody by virtue of its temperature.

1. **Kirchoff's Law :** *"The radiation law which states that at a given temperature the ratio of the emissivity to the absorptivity for a given wavelength is the same for all bodies and is equal to the emissivity of an ideal blackbody at that temperature and wavelength".*

Since no real body is a perfect emitter, its emittance is less than that of a black body. Thus, the emissivity (€) of a real (gray) body is defined by,

$€_g = M_g / M_b$

Where, $€_g$ is the emissivity of a real (grey) body

M_g is the emittance of a real (grey) body

M_b is the emittance of a black (grey) body (perfect radiator)

This radiative transfer law is very important, when considering energy balance. It states that at thermal equilibrium, the emissivity (õ) of a body or surface equals its absorptivity (±).

Mathematically, we can conceptualize Kirchoff's law as

EEb=*õ* = ±

The radiant energy emitted from a real surface is represented as E (*W*/*m*2), while that of a blackbody (a theoretical condition) is given by *EB* (*W*/*m*2).

Simply put, a surface at steady state temperature will absorb light equally as well as it emits light. Though light is directional, surfaces exchange photons in both directions.

2. **Planck's Law :** "*An expression for the variation of emittance of a blackbody at a particular temperature as a function of wavelength*". This law is generalized to mean that all objects have some internal temperature, and given that temperature, they all glow. Max Planck was able to establish the dependence of the spectral emissive energy of a blackbody for all wavelengths of light (*E»,b*), given a known equilibrium temperature of the black body.

$$E\lambda, b = 2\pi hc2\lambda 5\left[ehc\lambda KT - 1\right]$$

3. **Wien's Displacement Law :** "*Describes the shift of the radiant power peak to shorter wavelengths as temperature increases*".

The Wien's displacement law provides us with expected values for the most probable wavelengths in the Bose Einstein distribution of black body radiation. The law implies that the distribution of photons emitted from a surface at any temperature will have the same form or shape as a distribution of photons emitted from a surface at any other temperature. Mathematically,

$$\lambda \max = 2.8978x106nmKT(K)$$

4. **Stefan-Boltzmann Law :** "*Stefan - Boltzmann constant - 5.68 x 10^12 W.cm -2. K-4; appropriate to Law of same name stating that the amount of energy radiated per unit time from a unit surface of an ideal blackbody is proportional to the fourth power of the absolute temperature (in units of K) of that body*".

The Stefan-Boltzmann Law states that the radiative energy emitted by a surface is proportional to the fourth power of the surface's absolute temperature. The Stefan-Boltzmann Law shows that if you were to integrate all the energies from the wavelengths in Planck's Law, you have an analytical solution of the form below:

$Eb=\int\infty 0E\lambda,bd\lambda=\sigma T4$, where σ = Stefan-Boltzmann constant $=5.67\times10"8W/m2K4$

Keep in mind that this is for the **surface** of the emitter. A surface like the Sun will have a very high value for the energy density on the surface, which then decreases in proportion with the **Inverse Square Law** from the last page (over the 93 million miles distance to the Earth's surface).

Chapter 5

Aerial Photography and its Interpretation

(Concepts, active and passive remote sensing, equipment and materials-aerial bases, cameras, films, filters, stereoscopes, sensors and their characters)

Concepts of Aerial Photography

What is an aerial photograph?

An aerial photograph, in broad terms, is any photograph taken from the air. Normally, air photos are taken vertically from an aircraft using a highly accurate camera. There are several things you can look for to determine what makes one photograph different from another of the same area, including type of film, scale, and overlap. Other important concepts used in aerial photography are stereoscopic coverage, fiducial marks, focal length, roll and frame numbers, and flight lines and index maps. The following material will help you understand the fundamentals of aerial photography by explaining these basic technical concepts.

What information can I find on an air photo?

Unlike a map, features on an aerial photograph are not generalized or symbolized. Air photos record all visible features on the Earth's surface from an overhead perspective. Although the features are visible, they are not always easily identifiable. The process of studying and gathering the information required for the identification of the various cultural and natural features is called photo interpretation. With careful interpretation, air photos are an excellent source of spatial data for studying the Earth's environment.

Basic Concepts of Aerial Photography

Film : Most air photo missions are flown using black and white film, however, colour, infrared, and false-colour infrared film are sometimes used for special projects.

Focal length : The distance from the middle of the camera lens to the focal plane (i.e. the film). As focal length increases, image distortion decreases. The focal length is precisely measured, when the camera is calibrated.

Scale : The ratio of the distance between two points on a photo to the actual distance between the same two points on the ground (i.e. 1 unit on the photo equals "x" units on the ground). If a 1 km stretch of highway covers 4 cm on an air photo, the scale is calculated as follows:

$$\frac{\text{Photo Distance}}{\text{Ground Distance}} = \frac{4\,\text{cm}}{1\,\text{km}} = \frac{4\,\text{cm}}{100000\,\text{cm}} = \frac{1}{25000} \quad \text{Scale: } 1/25000$$

Another method used to determine the scale of a photo is to find the ratio between the camera's focal length and the plane's altitude above the ground being photographed.

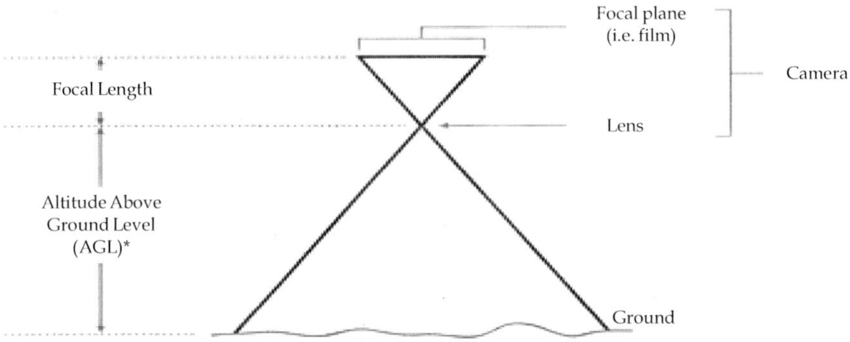

* Altitude Above Ground Level (AGL) is calculated by subtracting the ground elevation from the plane's altitude Above Sea Level (ASL).

If a camera's focal length is 152 mm, and the plane's altitude Above Ground Level (AGL) is 7 600 m, using the same equation as above, the scale would be:

$$\frac{\text{Focal Length}}{\text{Altitude(AGL)}} = \frac{152\,\text{mm}}{7600\,\text{m}} = \frac{152\,\text{mm}}{7600000\,\text{mm}} = \frac{1}{50000} \quad \text{Scale: } 1/50000$$

Scale may be expressed three ways :

a) Unit Equivalent

b) Representative Fraction

c) Ratio

A photographic scale of 1 millimetre on the photograph represents 25 metres on the ground would be expressed as follows:

- Unit Equivalent - 1 mm = 25 m
- Representative Fraction - 1/25 000
- Ratio - 1:25 000

Two terms that are normally mentioned, when discussing scale are:

1. **Large Scale** - Larger-scale photos (e.g. 1/25 000) cover small areas in greater detail. A large scale photo simply means that ground features are at a larger, more detailed size. The area of ground coverage that is seen on the photo is less than at smaller scales.

2. **Small Scale** - Smaller-scale photos (e.g. 1/50 000) cover large areas in less detail. A small scale photo simply means that ground features are at a smaller, less detailed size. The area of ground coverage that is seen on the photo is greater than at larger scales.

The National Air Photo Library has a variety of photographic scales available, such as 1/3000 (large scale) of selected areas, and 1/50000 (small scale).

Fiducial marks: Small registration marks exposed on the edges of a photograph. The distances between fiducial marks are precisely measured when a camera is calibrated, and cartographers, when compiling a topographic map use this information.

Overlap: This is the amount by which one photograph includes the area covered by another photograph, and is expressed as a percentage. The photo survey is designed to acquire 60 per cent forward overlap (between photos along the same flight line) and 30 per cent lateral overlap (between photos on adjacent flight lines).

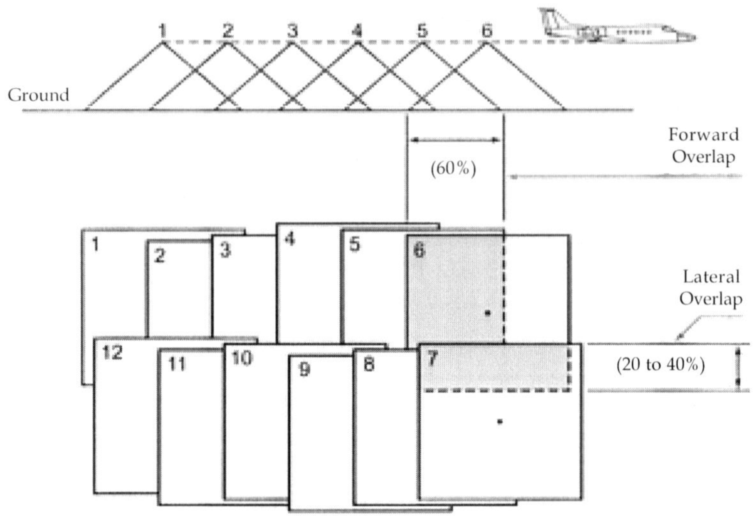

Photographic Overlap

Stereoscopic Coverage : The three-dimensional view, which results when two, overlapping photos (called a stereo pair), are viewed using a stereoscope. Each photograph of the stereo pair provides a slightly different view of the same area, which the brain combines and interprets as a 3-D view.

Roll and Photo Numbers : Each aerial photo is assigned a unique index number according to the photo's roll and frame. For example, photo A23822-35 is the 35th annotated photo on rolls A23822. This identifying number allows you to find the photo in NAPL's archive, along with metadata information, such as the date it was taken, the plane's altitude (above sea level), the focal length of the camera, and the weather conditions.

Flight Lines and Index Maps : At the end of a photo mission, the aerial survey contractor plots the location of the first, last, and every fifth photo centre, along with its roll and frame number, on a National Topographic System (NTS) map. Small circles represent photo centres, and straight lines are drawn connecting the circles to show photos on the same flight line.

This graphical representation is called an air photo index map, and it allows you to relate the photos to their geographical location. Small-scale photographs are indexed on 1/250 000 scale NTS map sheets, and larger-scale photographs are indexed on 1/50 000 scale NTS maps.

Mosaic : A **mosaic** is a photographic reproduction of a series of aerial photographs put together in such a way that the detail of one photograph matches the detail of all adjacent photographs. Mosaics are, for the most part, reproduced at a much smaller scale than the original photography, and consist of three main types: **uncontrolled mosaics, semi-controlled mosaics** and **controlled mosaics**.

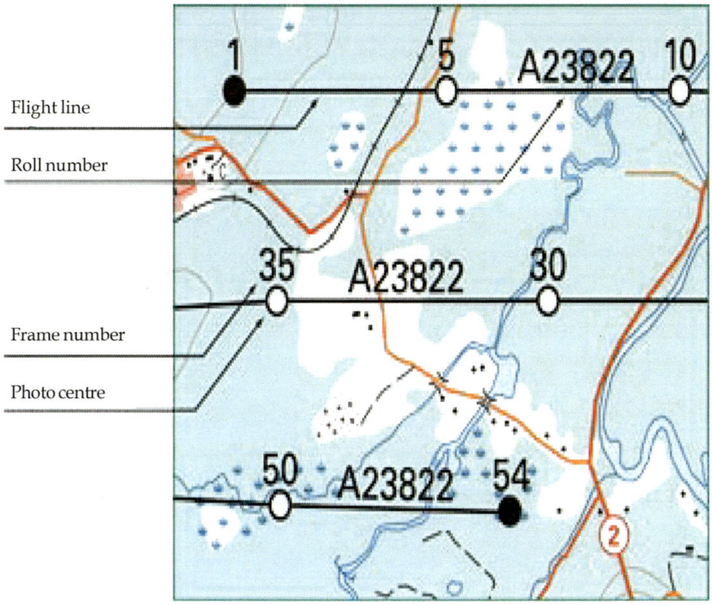

Uncontrolled Mosaics

- Prints are laid out so as to join together in a "best fit" scenario.
- Prints may be tone-matched.

Semi-Controlled Mosaics

- The prints used in the mosaic are tone-matched but not rectified.
- Prints are laid down to fit a map base of the same scale.
- A title and scale may be added.

Controlled Mosaics

- Clients must supply a map base as well as a minimum of three (3) ground control points per print.
- Prints are tone-matched and rectified to fit the map base.

Aerial Camera Lens and Aerial Photograph

1. Aerial Camera

A camera specially designed for use in aircraft. The camera is a light proof chamber or box in which the image of an exterior object is projected upon a sensitized plate or film through an opening, usually equipped with a lens, shutter and variable aperture. In India, the aerial cameras available are RC-5(a), RC-8 (universal), RC-10, RMK 'A', Eagle IX, I.I.S. (multiband), MK-70 and EL-500 types.

2. Photographic Lens

The essential fact of photography is that an image projected through a converging lens system is recorded on a sensitized surface upon which it can be developed and fixed.

The function of photography is that an image projected of objects on the negative plane, i.e. on the emulsion surface of film stretched in the focal plane.

Lens equation is: $1/u + 1/v = 1/f$

u is the distance of object from the lens, v is the distance of image from the lens and f is the focal length of the lens.

In an ordinary camera, image is brought to a clear focus by adjusting the negative plane, i.e. adjusting v, but in a surveying camera (used for air photography) u is very large compared to f (e.g. f = 6 inches, u = 30,000 ft) and may be assumed infinitely great compared to v, so that v equals f (approximately). The camera is, therefore, not provided with any focusing arrangement and hence, is a fixed focus camera. All images are formed in the focal plane which passes through the rear principal focal point F and is perpendicular to the optical axis (principal axis of the lens). This plane is called the negative plane or focal plane.

In aerial photography, the rays from the object space meet the principal axis at a point called Front Node N1 and leave it at Rear Node N2 towards the image space, parallel to their respective directions. That means, any ray which appears to converage at N1 appears to diverge from N2.

The Front Node N1 is often called External Perspective Centre and the Rear Node N2 the Internal Perspective Centre. The line joining these two points is the optical or principal axis of the lens system.

A single lens cannot produce a perfect image and the object point will not be images as a point but as a blur. Image imperfections are called aberration and are classified as:

- Spherical aberration
- Coma
- Chromatic aberration
- Astigmatism
- Curvature of the field
- Lens distortion

In order to have good image quality which is very essential for photogrammetry and photo-interpretation a combination of lenses of glasses of different optical qualities e.g. refractive indexes are used in a photographic lens to do away with or minimize the aberration effects. As such the nodes N1 and N2 may have an appreciable separation.

Now we may define focal length, the knowledge of which is of great importance in photogrammetry, as the distance of the internal perspective centre from the negative plane – the plane of best average definition.

Nodal Point Seperation

The flying height 'Z' above the terrain is so large that the separation of N1 and N2 is negligible compared to it and replaced by a point for the determination of Z (flying height), but for the determination of focal length this separation N1, N2 is significant.

3. Aerial Photograph

Types of Photographs

Photographs, which are used for mapping and photo-interpretation, can be divided into the following main classes according to the direction of the camera axis:

a) Vertical photographs
b) Horizontal or terrestrial photographs and
c) Oblique photographs.

The terms 'vertical' and 'horizontal' refer to the direction in which the camera axis was pointing at the time of exposure.

Vertical Air Photographs

These are taken with the axis of the aerial camera vertical or nearly vertical. A vertical photograph closely resembles a map and is particularly suitable for obtaining uniform coverage. As these photographs can be obtained with reasonably low tilt (tilt is deviation of the camera axis from the vertical) they are generally used for mapping and photo-interpretation work and also extension of control.

Terrestrial Photographs

These are taken with photo-theodolites from camera stations on the ground with the axis of the camera horizontal and they present the more familiar elevation view. These types of photographs are used for survey of structures and monuments of architectural or archaeological value. Terrestrial photographs taken with normal good cameras can also be of considerable use in supplementing photo-interpretation of vertical aerial photographs particularly so in geology and forestry, where study of a profile may be needed.

Oblique Photographs

Aerial photographs taken with the optical axis of the aerial camera tilted from the vertical are known as oblique photographs. These photographs cover large areas of ground but clarity of details diminishes towards the far end of the photographs. Aerial photographs on which the horizon does not appear are known as low oblique and are, sometimes, used to compile reconnaissance maps of inaccessible areas. High oblique photographs, which are tilted sufficiently to contain the horizon, were previously used for extension of planimetric and height control, when the available ground control was insufficient to provide necessary accuracy. These have very limited use at present.

There are combinations of above types of photography taken with two or more cameras in a single camera unit in the photographic air plane :

i) Convergent Photographs

These are low oblique photographs taken with two cameras exposed simultaneously at successive exposure stations, with their axes tilted at a fixed inclination from the vertical in opposite directions in the direction of the flight line so that the forward exposure of the first station forms stereo-pair with the backward exposure of the next station. Special plotting instruments are required for compiling topographical maps from convergent photographs.

ii)Trimetrogon Photography

Another type of photography which is a combination of a vertical and two oblique photographs is time trogon photography, in which the central photograph is vertical and the side ones are oblique. This photography can be used for rapid production of reconnaissance maps on small scales.

Cameras and Aerial Photography

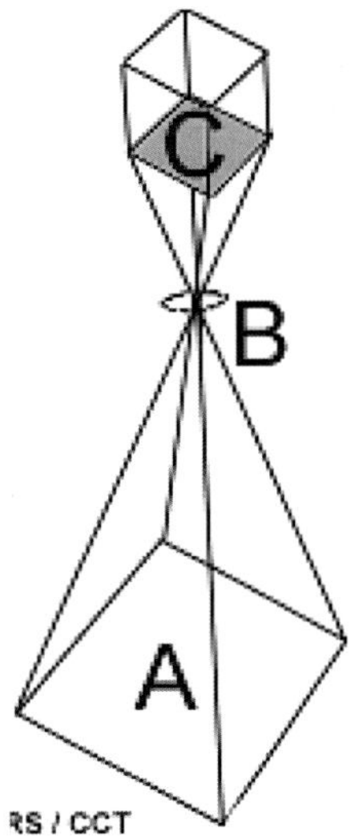

Cameras and their use for aerial photography are the simplest and oldest of sensors used for remote sensing of the Earth's surface. Cameras are **framing systems** which acquire a near-instantaneous "snapshot" of an **area (A)**, of the surface. Camera systems are passive optical sensors that use a **lens (B)** (or system of lenses collectively referred to as the optics) to form an image at the focal plane (C), the plane at which an image is sharply defined.

Photographic films are sensitive to light from 0.3 ìm to 0.9 ìm in wavelength covering the ultraviolet (UV), visible, and near-infrared (NIR). **Panchromatic** films are sensitive to the UV and the visible portions of the spectrum.

Panchromatic film produces black and white images and is the most common type of film used for aerial photography. UV photography also uses panchromatic film, but a filter is used with the camera to absorb and block the visible energy from reaching the film. As a result, only the UV reflectance from targets is recorded. UV photography is not widely used, because of the atmospheric scattering and absorption that occurs in this region of the spectrum. **Black and white infrared photography uses film sensitive to the entire 0.3 to 0.9 μm wavelength range** and is useful for detecting differences in vegetation cover, due to its sensitivity to IR reflectance.

Colour and false colour (or colour infrared, CIR) photography involves the use of a three layer film with each layer sensitive to different ranges of light.

Normal colour photograph

For a normal colour photograph, the layers are sensitive to blue, green, and red light - the same as our eyes. These photos appear to us the same way that our eyes see the environment, as the colours resemble those which would appear to us as "normal" (i.e. trees appear green, etc.). In colour infrared (CIR) photography, the three emulsion layers are sensitive to green, red, and the photographic portion of near-infrared radiation, which are processed to appear as blue, green, and red, respectively.

False colour photograph

In a false colour photograph, targets with high near-infrared reflectance appear red, those with a high red reflectance appear green, and those with a high green reflectance appear blue, thus giving us a "false" presentation of the targets relative to the colour we normally perceive them to be.

Most aerial photographs are classified as either **oblique** or **vertical**, depending on the orientation of the camera relative to the ground during acquisition.

- **Oblique aerial photographs** are taken with the camera pointed to the side of the aircraft. High oblique photographs usually include the horizon while low oblique photographs do not. Oblique photographs can be useful for covering very large areas in a single image and for depicting terrain relief and scale. However, they are not widely used for mapping as distortions in scale from the foreground to the background preclude easy measurements of distance, area, and elevation.
- **Vertical photographs** taken with a single-lens frame camera is the most common use of aerial photography for remote sensing and mapping purposes. These cameras are specifically built for capturing a rapid sequence of photographs while limiting geometric distortion. They are often linked with navigation systems onboard the aircraft platform, to allow for accurate geographic coordinates to be instantly assigned to each photograph. Most camera systems also include mechanisms which compensate for the effect of the aircraft motion relative to the ground, in order to limit distortion as much as possible.

Application of Cameras

1. Cameras can be used on a variety of platforms including ground-based stages, helicopters, aircraft, and spacecraft.
2. Very detailed photographs taken from aircraft are useful for many applications where identification of detail or small targets is required. The ground coverage of a photo depends on several factors, including the focal length of the lens, the platform altitude, and the format and size of the film.
3. The focal length effectively controls the **angular field of view** of the lens and determines the area "seen" by the camera. Typical focal lengths used are 90mm, 210mm, and most commonly, 152mm. The longer the focal length, the smaller the area covered on the ground, but with greater detail (i.e. larger scale).
4. The area covered also depends on the altitude of the platform. At high altitudes, a camera will "see" a larger area on the ground than at lower altitudes, but with reduced detail (i.e. smaller scale).
5. Aerial photos can provide fine detail down to spatial resolutions of less than 50 cm. A photo's exact spatial resolution varies as a complex function of many factors which vary with each acquisition of data.

When obtaining vertical aerial photographs, the aircraft normally flies in a series of lines, each called a **flight line**. Photos are taken in rapid succession looking straight down at the ground, often with a **50-60 percent overlap (A) between successive photos** (fig). The overlap ensures total coverage along a flight line and also facilitates **stereoscopic viewing**. Successive photo pairs display the overlap region from different perspectives and can be viewed through a device called a **stereoscope** to see a three-dimensional view of the area, called a **stereo model**. Many applications of aerial photography use stereoscopic coverage and stereo viewing.

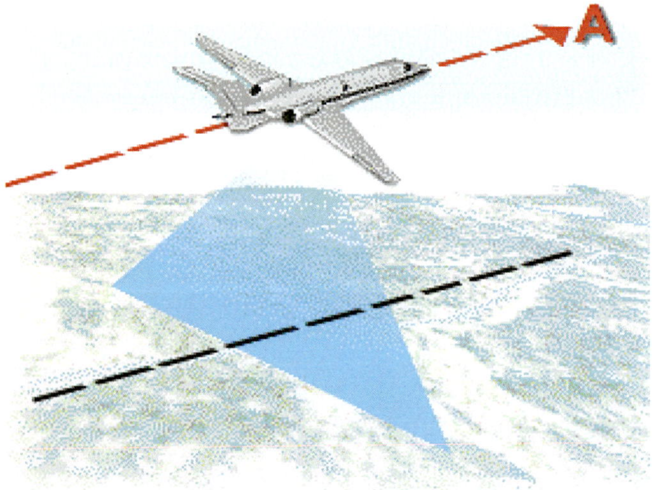

Aerial photographs are most useful, when fine spatial detail is more critical than spectral information, as their spectral resolution is generally coarse, when compared to data captured with electronic sensing devices. The geometry of vertical photographs is well understood and it is possible to make very accurate measurements from them, for a variety of different applications (geology, forestry, mapping, etc.).

The science of making measurements from photographs is called photogrammetry and has been performed extensively since the very beginnings of aerial photography. Photos are most often interpreted manually by a human analyst (often viewed stereoscopically). They can also be scanned to create a digital image and then analyzed in a digital computer environment. In Chapter 4, we will discuss in greater detail, various methods (manually and by computer) for interpreting different types of remote sensing images.

Multiband photography

Multiband photography uses multi-lens systems with different film-filter combinations to acquire photos simultaneously in a number of different spectral ranges. The advantage of these types of cameras is their ability to record reflected energy separately in discrete wavelength ranges, thus providing potentially better separation and identification of various features. However, simultaneous analysis of these multiple photographs can be problematic.

Digital cameras

Digital cameras, which record electromagnetic radiation electronically, differ significantly from their counterparts which use film. Instead of using film, digital cameras use a gridded array of silicon coated CCDs (charge-coupled devices) that individually respond to electromagnetic radiation. Energy reaching the surface of the CCDs causes the generation of an electronic charge which is proportional in magnitude to the "brightness" of the ground area. A digital number for each spectral band is assigned to each pixel based on the magnitude of the electronic charge. The digital format of the output image is amendable to digital analysis and archiving in a computer environment, as well as output as a hardcopy product similar to regular photos. Digital cameras also provide quicker turn-around for acquisition and retrieval of data and allow greater control of the spectral resolution. Although parameters vary, digital imaging systems are capable of collecting data with a spatial resolution of 0.3m, and with a spectral resolution of 0.012 mm to 0.3 mm. The size of the pixel arrays varies between systems, but typically ranges between 512 x 512 to 2048 x 2048.

Elements, Aids, Techniques, Methods & Procedures of Airphoto Interpretation

I. Definitions

Photo Interpretation: The act of examining aerial photographs/ images for the purpose of identifying objects and judging their significance.

Photography: The art or process of producing images on a sensitized surface by the action of light or other radiant energy.

Image: A reproduction or imitation of the form of a person or thing. The optical counterpart of an object produced by a lens or mirror or other optical system.

Photogrammetry: The science or art of obtaining reliable measurements by means of photography.

Before an interpreter commences the reading of the photo, the concept of stereoscopy must be understood.

Stereoscopes

The function of a stereoscope is to deflect normally converging lines-of-sight so that each eye views a different photographic image.

Stereoscopes are grouped into 2 basic types:

i) Lens stereoscopes

ii) Mirror prism stereoscopes

Pocket Stereoscope

By far the most popular is the lens stereoscope commonly known as pocket stereoscope. The pocket stereoscope usually has plane-convex lens, upper side flat with a focal length of 100 mm. The rays entering the eyes are now parallel and converge at infinity and have been accommodated (focused) at 100 mm distance (Fig. *15*). Since the normal viewing distance is 250 mm, a closer view, i.e. at 100 mm results in a magnification. The magnification is then 250/100 = 2.5. More expensive types have a changeable eye-base. Such a refinement is not necessary for operators with an average eye-base range of 60 to 68 mm. The pocket stereoscope is cheap, transportable, and has a large field of view.

It has two big disadvantages

a) Limited magnification. Pocket stereoscopes with more than three times magnification cannot be equipped with simple plane-convex lenses, due to the too large an increase in lens aberrations. In addition, the distance between the head and the photos becomes too small for adequate illumination without undue complications.

b) The distance between corresponding points on the photos must be equal to or smaller than the eye-base. With normal size photographs, this becomes difficult or impossible without bending or folding the photos.

It should not be forgotten, however, that due to the simple optical system the image quality of the pocket stereoscope is very good.

Mirror Stereoscope

The two above mentioned drawbacks have led to the development of the mirror stereoscope. The normal size photos (23 cm x 23 cm) can be separated and seen under the stereoscope without folding them. The path of the bundle of rays has been diverted and brought to the eyes at 65 mm separation. This is achieved by reflecting mirrors. Normally the distance between corresponding points is kept at 240 mm so that photographs are placed separately, i.e., it effectively increases the eye base from 65 mm to 240 mm. As in pocket stereoscope the picture must be at the focal plane of the lenses in order to have convergence at infinity. The mirrors M 1 are placed in such a way that the picture distance via the small mirrors M2 (generally prisms) become equal to the focal length of the lens, usually 300 mm. This gives approximately 250/300 = 0.8 x magnification, or rather reduction the picture observed, to magnify the image additional oculars of magnification 3x to 8x can be used over the prisms or a lens placed before each prism giving a magnification of about 1.8x.

Stereoscopy

A pair of stereoscopic photographs or images can be viewed stereoscopically by looking at the left image with the left eye and the right image with the right eye. This is called **stereoscopy**. Stereoscopy is based on Porro-Koppe's Principle that the same light path will be generated through an optical system, if a light source is projected onto the image taken by an optical system. The principle will be realized in a **stereo model,** if a pair of stereoscopic images is re-constructed using the relative location or tilt at the time the photography was taken. Such an adjustment is called relative orientation in photogrammetric terms. The eye-base and the photo-base must be parallel in order to view at a stereoscopic model, as shown in the Fig. 3.1.

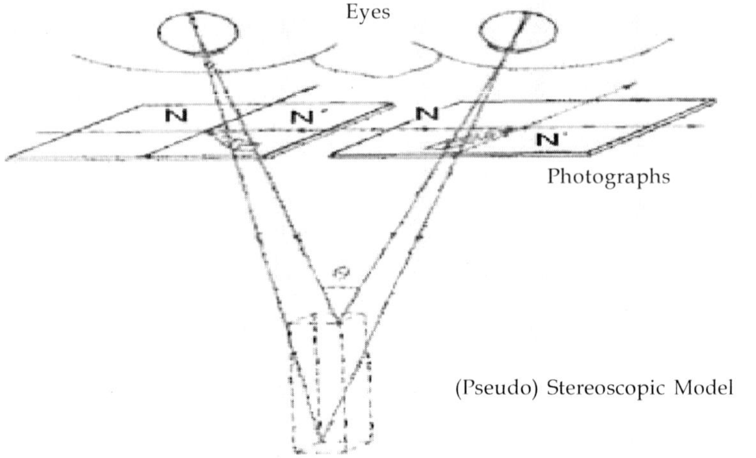

Fig. 3.2: Schematic diagram of stereoscopic vision

Usually a stereoscope is used for image interpretation. There are several types of stereoscope, for example, portable lens stereoscope, stereo mirror scope (see Fig. 3.1) stereo zoom transfer scope, etc.

Fig. 3.2: Stereoscopic viewer

The process of stereoscopy for aerial photographs is as follows: At first the centre of both aerial photographs, called the **principal point**, should be marked. Secondly, the principal point of the right image should be plotted in its position on the left image. At the same time the principal point of the left image should be also plotted on the right image. These principal points and transferred points should be aligned along a straight line, called the base line, with an appropriate separation (normally 25-30 cm in the case of a stereo mirror scope as shown in Fig. 3.3. By viewing through the binoculars a stereoscopic model can now be seen.

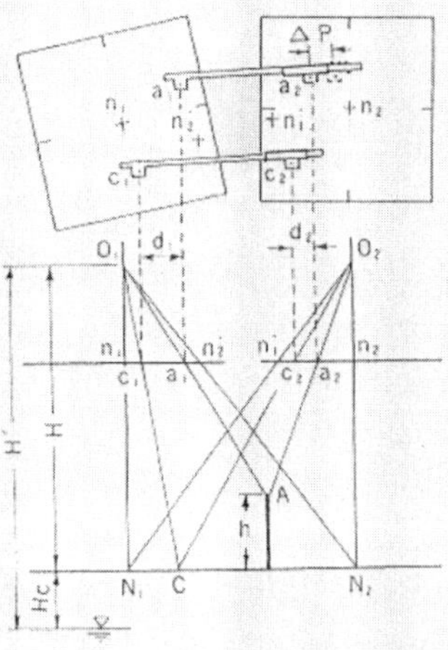

Hc : flight height

c : control point

Hc : height of control point

H : H' - Hc

$\overline{n_1 \cdot n_2}$: principal points on a pair of stereo photos

$\overline{n_1 \cdot n_2}$: transferred points from each principal point of the photos

$\overline{n_1 \cdot n_2}$: principal distance

$\overline{c_1 \cdot c_2}$: distance between at point c on a pair of stereo photos

$\overline{a_1 \cdot a_2}$: distance between at point a on a pair of stereo photos

parallax at point c : $P = \overline{n_1 \cdot n_2} - \overline{c_1 \cdot c_2}$

parallax distance between point a & c : $\Delta P = \overline{c_1 \cdot c_2} - \overline{a_1 \cdot a_2}$

obtained height : $h = H \cdot \Delta P / (P + \Delta P)$

Fig. 3.3: Principle of height measurement

The advantage of stereoscopy is the ability to extract three dimensional information, for example, classification between tall trees and low trees, terrestrial features, such as height of terraces, slope gradient, detailed geomorphology in flood plains, dip of geological layers and so on.

The principle of height measurement by stereoscopic vision is based on the use of **parallax**, which corresponds to the distance between image points, of the same object on the ground, on the left and right image. The height difference can be computed, if the **parallax difference** is measured between two points of different height, using a parallax bar, as shown in Fig. 3.3 above.

II. Activities of Airphoto/Image Interpretation

Detection/Identification - This is primarily a stimulus and response activity. The stimuli are the elements of image interpretation. The interpreter conveys his or her response to these stimuli with descriptions and labels that are expressed in qualitative terms e.g. likely, possible, or probable. Very rarely do interpreter use definite statements with regard to describing features identified on aerial photography.

Measurement - As opposed to detection and identification, the making of measurements is primarily quantitative. Techniques used by air photo interpreters typically are not as precise as those employed by photogrammetrists who use sophisticated instruments in making their measurements. Measurements made by photo interpreters will get you close; given high quality, high resolution, large-scale aerial photographs and appropriate interpretation tools and equipment, you can expect to be within feet; whereas with photogrammetry, if you employ the same type of photography and the appropriate equipment, you could expect to be within inches.

Problem Solving - Interpreters are often required to identify objects from a study of associated objects that they can identify; or to identify object complexes from an analysis of their component objects. Analysts may also be asked to examine an image, which depicts the effects of some process, and suggest a possible or probable cause. A solution may not always consist of a positive identification. The answer may be expressed as a number of likely scenarios with statements of probability of correctness attached by the interpreter.

Air photo interpretation is to photogrammetry as statistics is to mathematics; one deals with precision the other with probability.

III. Elements of Airphoto/Image Interpretation

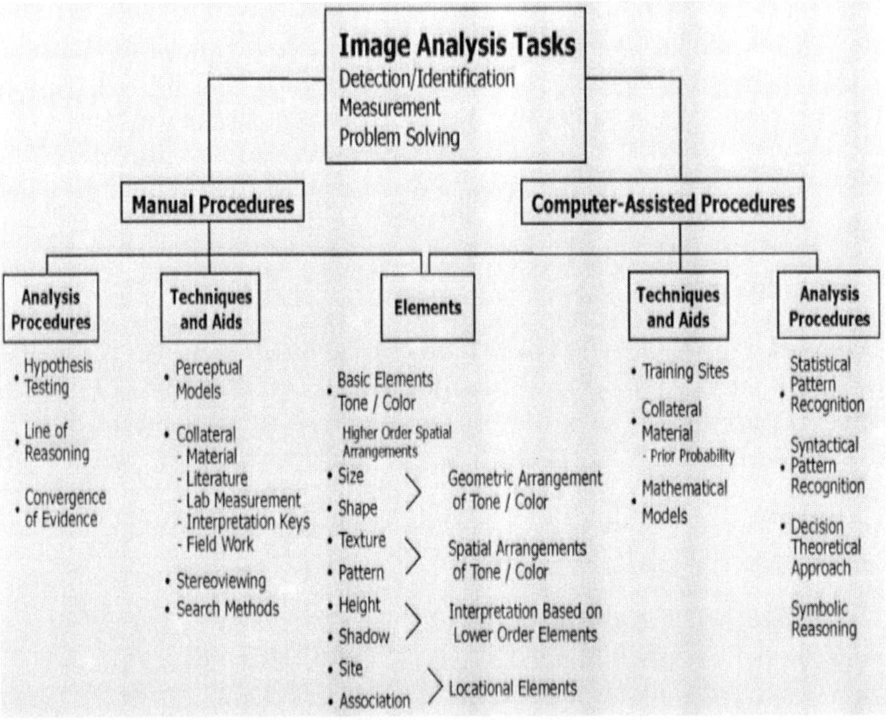

Primary Ordering of Image Elements Fundamental to the Analysis Process

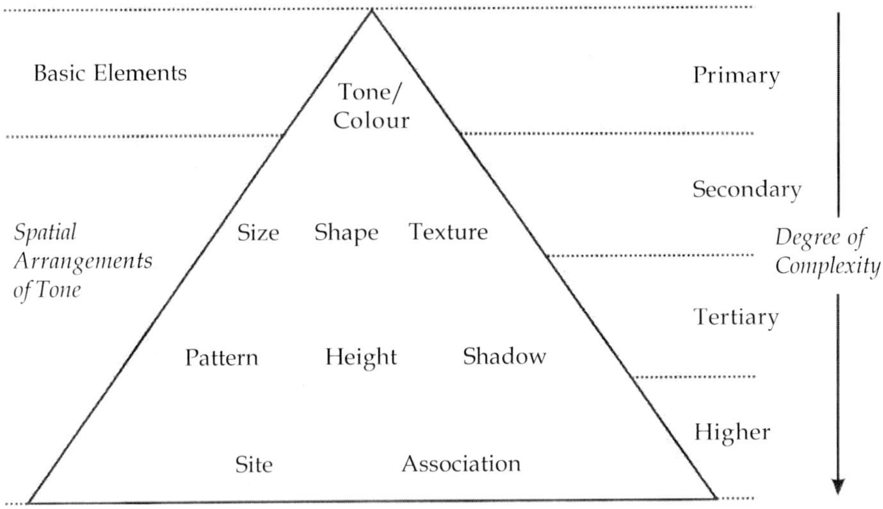

A. Basic - 1st Order

1. ***Tone/Colour***: Tone can be defined as each distinguishable variation from white to black. Colour may be defined as each distinguishable variation on an image produced by a multitude of combinations of hue, value and chroma. Many factors influence the tone or colour of objects or features recorded on photographic emulsions. But, if there is not sufficient contrast between an object and it's background to permit, at least, detection there can be no identification. While a human interpreter may only be able to distinguish between ten and twenty shades of grey; interpreters can distinguish many more colours. Some authors state that interpreters can distinguish at least 100 times more variations of colour on colour photography than shades of grey on black and white photography.

2. ***Resolution:*** Resolution can be defined as the ability of the entire photographic system, including lens, exposure, and processing, and other factors, to render a sharply defined image. An object or feature must be resolved in order to be detected and/or identified. Resolution is one of the most difficult concepts to address in image analysis because it can be described for systems in terms of modulation transfer (or point spread) functions, or it can be discussed for camera lenses in terms of being able to resolve

so many line pairs per millimeter. There are resolution targets that help to determine this, when testing camera lenses for metric quality. Photo interpreters often talk about resolution in terms of ground resolved distance, which is the smallest normal contrast object that can be identified and measured.

B. 2nd Order - Geometric Arrangements of Tone/Color

1. ***Size:*** Size can be important in discriminating objects and features (cars vs. trucks or buses, single family vs. multifamily residences, brush vs. trees, etc.). In the use of size as a diagnostic characteristic both the relative and absolute sizes of objects can be important. Size can also be used in judging the significance of objects and features. The size of the crowns of trees can be related to board feet that may be cut for specific species in managed forests. The size of agricultural fields can be related to water use in arid areas, or the amount of fertilizers used. The size of runways gives an indication of the types of aircraft that can be accommodated.

2. ***Shape:*** The shape of objects/features can provide diagnostic clues that aid identification. The Pentagon building in Washington is a diagnostic shape. For example, man-made features have straight edges while natural features tend not to. Roads can have right angle (90°) turns, railroads can't. Other examples include freeway interchanges, old fortifications (European cities), and military installations (surface to air missile sites).

C. 2nd. Order - Spatial Arrangement of Tone/Colour

1. ***Texture:*** Texture is the frequency of change and arrangement of tones. This is a micro image characteristic. The visual impression of smoothness or roughness of an area can often be a valuable clue in image interpretation. Still water bodies are typically fine textured, grass medium, brush rough. There are always exceptions though and scale can and does play a role; grass could be smooth, brush medium and forest rough on higher altitude aerial photograph of the same area.

2. ***Pattern:*** Pattern is the spatial arrangement of objects. Patterns can be either man-made or natural. Pattern is a macro image characteristic. It is the regular arrangement of objects that can be diagnostic of features on the landscape. An orchard has a

particular pattern. Pattern can also be important in geologic or geomorphologic analysis; drainage pattern can reveal a great deal about the lithology and geologic structural patterns of the underlying strata. Dendridic drainage patterns develop on flat-bedded sediments, radial on domes, linear or trellis in areas with faults, etc. It must be noted here that pattern is highly scale dependent.

D. 3rd. Order - Locational or Positional Elements

1. *Site:* Site refers to how objects are arranged with respect to one another, or with respect to terrain features. Aspect, topography, geology, soil, vegetation and cultural features on the landscape are distinctive factors that the interpreter should be aware of when examining a site. The relative importance of each of these factors will vary with local conditions, but all are important. Just as some vegetation grows in swamps others grow on sandy ridges or on the sunny side vs. the shaded sides of hills. Crop types may prefer certain conditions (e.g. orchards on hillsides). Man made features may also be found on rivers (e.g. power plant) or on hilltops (e.g. observatory or radar facility).
2. ***Association:*** Some objects are so commonly associated with one another that identification of one tends to indicate or confirm the existence of another. Smoke stacks, cooling ponds, transformer yards, coal piles, railroad tracks = coal fired power plant. Arid terrain, basin bottom location, highly reflective surface, sparse vegetation = playa, which typically have halophytic vegetation e.g. saltbush. Association is one of the most helpful interpretation clues in identifying man made installations. Aluminum manufacture requires large amounts of electrical energy. Schools of different grade levels typically have characteristic playing fields, parking lots and clusters of buildings. Nuclear power plants are associated with a source of cooling water; weather patterns can be associated with pollution sources, etc.

E. 3rd. Order - Interpreted from lower order elements

1. ***Height:*** For some types of analysis e.g. land forms, forestry and some intelligence applications, some interpreters believe that after tone/colour height is the most important element for identification. This is a point of debate, but height can add significant information in many types of interpretation tasks,

particularly those that deal with the analysis of man-made features and vegetation. How tall a tree can tell something about the expected amount of board feet? How deep an excavation can tell something about the amount of material that was removed (in some mining operations excavators are paid on the basis of material removed as determined by photogrammetric measurement of volume)?

2. *Shadow:* Geologists like low sun angle photography because of the features that shadow patterns can help identify (e.g. fault lines and fracture patterns). Church steeples and smokestacks can cast shadows that can facilitate their identification. Tree identification can be aided by an examination of the shadows thrown. Shadows can also inhibit interpretation. On infrared aerial photography shadows are typically very black and can render targets in shadows uninterpretable.

IV. Techniques of Photographic/Image Interpretation

Collateral Material

A review of all existing source material that pertains to a given area, process, type of facility or object, can aid in the interpretation process. The use of collateral material may also result in a better definition of the scope, objectives and problems associated with a given project. Also called "ancillary data", collateral material may come in the form of text, tables, maps, graphs, or image metadata. Census data, a map or description of the flora of a given area, a land use map, meteorological statistics, or agricultural crop reports can all be used in support of a given interpretation. Basically, collateral material represents data/information that an interpreter may use to aid in the interpretation process. Material contained within a Geographic Information System (GIS) that is used to assist an interpreter in an analysis task can be considered collateral data. Two classes of collateral materials deserve special mention: interpretation keys and field verification.

Interpretation Keys

An interpretation key is a set of guidelines used to assist interpreters in rapidly identifying features. Determination of the type of key and the method of presentation to be employed will depend upon,

a) The number of objects or conditions to be identified; and,

b) The variability typically encountered within each class of features or objects within the key.

Some authors say that as a general rule, keys are more easily constructed and used for the identification of man-made objects and features than for natural vegetation and landforms. For analysis of natural features, training and field experience are often essential to achieve consistent results. Basically, an interpretation key helps the interpreter organize the information present in image form and guides him/her to the correct identification of unknown objects. Keys can be used in conjunction with any type of remotely sensed data. Such keys can differ from those employed in other disciplines in that they can consist largely of illustrations, e.g. landforms, industrial facilities, military installations. Many types of keys are already available, if you can find or get your hands on them. This can often be very difficult and a reason why people develop their own keys.

Depending upon the manner in which the diagnostic features are organized, two types of keys are generally recognized :

1. Selective keys, and
2. Elimination keys.

Selective keys are arranged in such a way that an interpreter simply selects that example that most closely corresponds to the object they are trying to identify, e.g. industries, landforms, etc. Elimination Keys are arranged so that the interpreter follows a precise step-wise process that leads to the elimination of all items except the one(s) that he is trying to identify. Dichotomous keys are essentially a class of elimination key. Most interpreters prefer to use elimination keys in their analyses.

Field Verification

Field verification can be considered a form of collateral material because it is typically conducted to assist in the analysis process. Essentially, this is the process of familiarizing the interpreter with the area or type of feature. This type of verification is done prior to the interpretation to develop a visual "signature" of how the feature(s) of interest appear on the ground. After an interpretation is made, field verification can be conducted to verify accuracy. Fieldwork is sometimes calculated as being three times as expensive as lab analysis. (This is why good interpreters can be so valuable). The nature, amount, timing, method of acquisition, and data integration procedures should be carefully thought out. Will you use windshield surveys, point or transect sampling? Will the sampling be random or systematic?

The amount and type of field work required for a given project may vary greatly and is generally dependent upon the :

a) Type of analysis involved.

b) Image quality, including scale resolution and information to be interpreted.

c) Accuracy requirements for both classification, and boundary delineation.

d) Experience of the interpreter and the knowledge of the sensor, area, and subject.

e) Terrain conditions, and the accessibility of the study area.

f) Personnel availability, access to ancillary material.

g) Cost considerations.

Handling of Imagery

Although a good deal of photo interpretation is still done using paper prints, the use of diapositive transparencies is increasing. Transparencies can be used either as single frames or as a roll. Care should be taken when handling transparencies so that they are not marred. An orderly procedure for the handling of either prints or transparencies should be developed and adhered to in any interpretation project. Airphotos are typically numbered with flight name and/or frame number, and should be kept in order in so far as practical. Different dates and flight lines should be kept separate, etc. Anytime transparencies are used surfaces should be as clean as possible and the interpreter should either wear cotton gloves or be sure not to touch the emulsion surface as skin oils can cause image deterioration.

Stereo Viewing

Binocular vision is natural to all of us, but to the trained interpreter the ability to perceive stereo is an incredibly valuable asset. Stereo viewing will be covered in detail later, but suffice it to say that viewing high quality stereo aerial photography though a mirror stereoscope is like seeing in another dimension. Although the identification and interpretation of many landscapes can be accomplished with mono, stereo is required for certain types of studies. The following are some tips for using stereo effectively.

Basics for Stereo viewing

1. Make certain that the photos are properly aligned, preferably with the shadows falling toward the viewer.
2. Keep the eye base and the long axis of the stereoscope parallel to the flight line.
3. Maintain an even glare free illumination on the prints or transparencies.
4. Arrange for comfortable sitting and sufficient illumination.
5. Keep the lenses of the stereoscope clean, properly focused and separated to your interpupillary distance.
6. The novice interpreter should not work with stereo more than 30 minutes out of any hour period. You have not had a headache until you've had one that comes from doing stereo interpretation for too long.

Trouble Shooting Stereo

1. Your eyes may be of unequal strength. If you normally wear glasses for reading or close-up work, you should also wear glasses when using the stereoscope.
2. Poorly illumination, misaligned prints or uncomfortable viewing positions may result in eye fatigue.
3. Illness or severe emotional distress may create sensations of dizziness in one using a stereoscope.
4. Reversal of prints may cause psuedo-stereo. A similar problem may occurs if prints are aligned with the shadows falling away from rather than towards the interpreter.

5. Objects that change positions between exposures cannot be viewed in stereo.
6. In areas of steep topography, scale differences in adjacent photographs may make it difficult to obtain a three dimensional image.
7. Dark shadows or clouds may prohibit stereo viewing of an area by obscuring an object on one photo.

D. Further Definitions: The Multi Concept

Multi-Station: The successive overlapping of images taken along a given flight line as being flown by an aircraft or by a satellite along an orbit path. Think of it like "multi-position". Not to be confused with multi-stage.

Multi-Band: Multi-band indicates individual spectral bands within a given region of the EM spectrum (e.g. the red green and blue bands of the visible portion of the EM spectrum). Often seen to have an overlapping meaning with the next term, multi-spectral.

Multi-Spectral: The use of images from various regions of the EM spectrum (e.g. ultra-violet, visible, infrared, thermal and microwave).

Multi-Date: The use of multiple aerial photographs or remotely sensed images taken over time of a given area.

Multi-Stage: This typically means using ground based photos, oblique low altitude photos and vertical photographs or remotely sensed images from platforms flying at different altitudes. Multi-stage has also been applied to sampling strategies; A multi-stage sampling scheme as used in statistics is one where progressively more information is obtained for progressively smaller sub-samples of the area being studied.

Multi-Direction: There are times when more information can be obtained using viewing angles other than vertical.

Multi-Disciplinary: Basically, no one interpreter can know everything about a system in question. By using teams of interpreters and experts with expertise in different disciplines more information may be gained for a given application. In the legal system, this is most similar to the "convergence of evidence" idea; having different viewpoints and different information sources to prove a point adds validity.

Multi-Thematic: Remote sensing images are one-time write, many times read. Many different themes (e.g. hydrology, vegetation, transportation, urban areas, etc.) can be extracted from a single set of images.

Multi-Use: Many potential users from environmental planners to resource managers to public policy decision-makers can use outputs derived from image analysis and interpretation.

E. Methods of Search

There are basically two techniques that people tend to follow, when searching for imagery. One is logical search and the other can be termed the "fishing expedition". In the latter, the fishing expedition, the interpreter searches the imagery in a random fashion attempting to find recognizable features or object that will lead to whatever the interpretation goal happens to be. At some point even this type of interpretation begins to logically converge. Patterns of anomalous vegetation may lead to looking for water sources, which may lead to looking for transportation systems, illegal growing, etc. Logical search is a more systematic method of analysis most often used by interpreters.

The logical search involve these things :

1. The interpreter should always keep in mind the basic qualities of the imagery they are dealing with, e.g. film filter combination, the season and time of day of acquisition, and the image scale, etc. In addition the interpreter should always remember to examine all the titling information on an image.
2. Interpretation should begin with the general and proceed to the specific. After gaining an overall impression of the photograph the interpreter should begin to examine the physical features (e.g. water bodies, mountains, forests, etc.) and cultural features (e.g. urban areas, farms, road networks, etc.). The interpreter should then move to more specific questions e.g. what type of trees makes up the forest? What types of roads are present?
3. Interpretation should be conducted logically one step at a time. Following from 2 above it is good to go from a detailed examination of landforms to vegetation, to hydrology and so on.

Then address cultural features in the same fashion. What types of urban features are present in single-family residences, multi-family residences, industries, retail districts and so on?

F. Convergence of Evidence

1. Image interpretation is basically a deductive process. Features that can be detected and identified lead the interpreter to the location and identification of other features. This is convergence, and for many applications of air photo interpretation, this involves the activities of one or two individuals synthesizing a large amount of information.
2. Deductive interpretation requires either the conscious or unconscious consideration of all of the elements of image interpretation. The completeness and accuracy of an interpretation is in some measure proportional to the interpreters understanding of the "how and the why" of the elements, techniques and methods of interpretation.

❑❑❑

Chapter 6

Photogrammetry

Photogrammetry is an estimative scientific method that aims at recovering the exact positions and motion pathways of designated reference points located on any moving object, on its components and in the immediately adjacent environment. Photogrammetry employs high-speed imaging and the accurate methods of remote sensing in order to detect, measure and record complex 2-D and 3-D motion fields (see also SONAR, RADAR, LiDAR, etc.). Photogrammetry feeds the measurements from remote sensing and the results of imagery analysis into computational models in an attempt to successively estimate, with increasing accuracy, the actual, 3-D relative motions within the researched field.

Its **applications** include satellite tracking of the relative positioning alterations in all Earth environments (e.g. tectonic motionsm etc.), the research on the swimming of fish, of bird or insect flight, other relative motion processes (International Society for Photogrammetry and Remote Sensing). The quantitative results of photogrammetry are then used to guide and match the results of computational models of the natural systems, thus helping to invalidate or confirm new theories, to design novel vehicles or new methods for predicting or/and controlling the consequences of earthquakes, tsunamis, any other weather types, or used to understand the flow of fluids next to solid structures and many other processes.

Photogrammetry is as old as modern photography, can be dated to the mid-nineteenth century, and its detection component has been emerging from radiolocation, multilateration and radiometry while its 3-D positioning estimative component (based on modeling) employs methods related to triangulation, trilateration and multi-dimensional scaling.

In the simplest example, the distance between two points that lie on a plane parallel to the photographic image plane can be determined by measuring their distance on the image, if the scale (s) of the image is known. This is done by multiplying the measured distance by $1/s$.

Algorithms for photogrammetry typically attempt to minimize the sum of the squares of errors over the coordinates and relative displacements of the reference points. This minimization is known as bundle adjustment and is often performed using the Levenberg–Marquardt algorithm.

Definitions

Photogrammetry : The science or art of obtaining reliable measurements by means of photographs (taken from calibrated cameras).

Photo-interpretation: The art of examining photo-images for the purpose of identifying objects and judging their significance.

Flying height: The vertical distance of the aircraft in flight above a given datum (usually mean sea-level).

Datum plane: A reference plane in relation to which the positions of other points are determined, e.g. heights are reckoned from the mean sea, level. Hence, mean sea, level is the datum plane for height.

Exposure station: The point in space or on the ground occupied by the camera lens at the moment of exposure. This is also called camera station. In the case of aerial photography, it is also called air station.

Exposure interval: The time interval between two successive exposures.

Exposure time: The time during which the light sensitive material is subjected to the action of light, i.e. the time during which the camera shutter is open to allow the light to enter the camera.

Air base: The line joining two successive camera stations of aerial photography or its length.

Negative: A photographic image on film or plate in which the subject tones are reversed, i.e. dark appears light and light appears dark or complementary it colours are concerned.

Tone: Each distinguishable shade variation from black to white.

Positive: A photographic image having approximately the same rendering of tones as the original subject photographed.

Diapositive: A positive photographs on a transparent medium is called diapositive.

Contact print: A print made from a negative or a diapositive in direct contact with sensitized material.

Fiducial marks: These are index marks, usually four which are rigidly connected with the camera lens through the camera body and which form images on the negative, when exposure is made. This set of four fiducial marks defines the principal point of the photographs in an adjusted camera.

Fiducial axis: The ling joining opposite fiducial marks on a photograph.

Corresponding images: The images of the same object on two or more photographs.

Overlays: Amount by which one photograph overlaps the area covered by another. It is expressed as a percentage. The overlaps between aerial photographs in the same strip are called forest and aft overlap or forward lap and end lap. The overlap between photographs of adjoining strips is called Lateral overlap or side lap.

Strip: Any number of overlapping photographs taken along a photo-flight line, usually at an approximately constant altitude.

Photogrammetry may be divided into two classes :

- Terrestrial or ground photogrammetry
- Aerial photogrammetry

1. Terrestrial or Ground Photogrammetry

In terrestrial photogrammetry maps are prepared from terrestrial (or ground) photographs or terrestrial photogrammetry employees photograph taken form points on earth surface for measurement purposes.

The terrestrial photographic surveying considered as the further development of plane table surveying.

2. Aerial Photogrammetry

In Aerial photogrammetry, maps ae produced from air photograph's (photographs taken from the air).

Photogrammetry in compassed two major areas of specialization :

- Metrical
- Interpretive

The first area is of principal interest to surveyor's since it is applied to determined dittanies elevations areas volumes cress-section's an to compile topographic maps form measurement made on photographs.

Interpretive photogrammetry involves objects from there photographic images and judging their significances. Critical factors considered in identifying objects of shape, sizes, patterns, shadow.

Uses of Photogrammetry

Photography dates to 1839

- Photogrammetry is used in preparing topographic maps and its advantages include :
  - Speed of coverage of an area
  - Relatively low cost
  - Ease of obtaining topographic details
  - Reduce likely wood of omitting date due to tremendous amount of detail shown in photographs
- Photogrammetry is used in land surveying to compute co-ordinates of section corner, body corner or points of **Photo Theodolite.**

Photo theodolite

A photo theodolite is a form of ground camera. It is a combination of camera and theologize and is used of taking photographic and measuring the angles which the rival plane of collimation makes with base line.

- The optical axis of the camera is arranged parallel to the line of collimation of the chocolate telescope and the theologize allows the determination of the orientation of the camera axis.
- The angular measurement is needed to fix the camera stations (A and B) during the control survey.

Photo theodolite

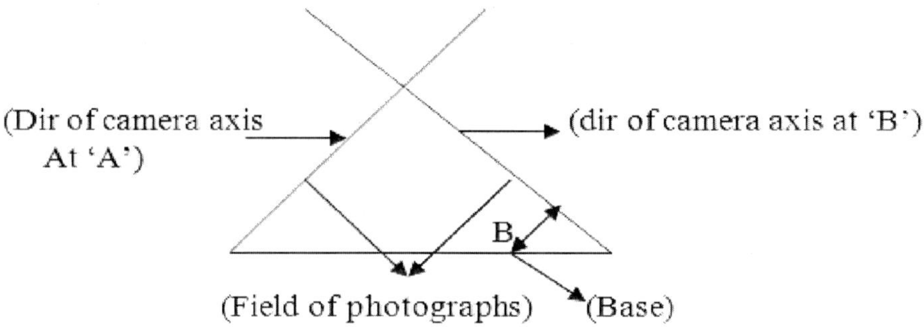

Photogrammetric methods

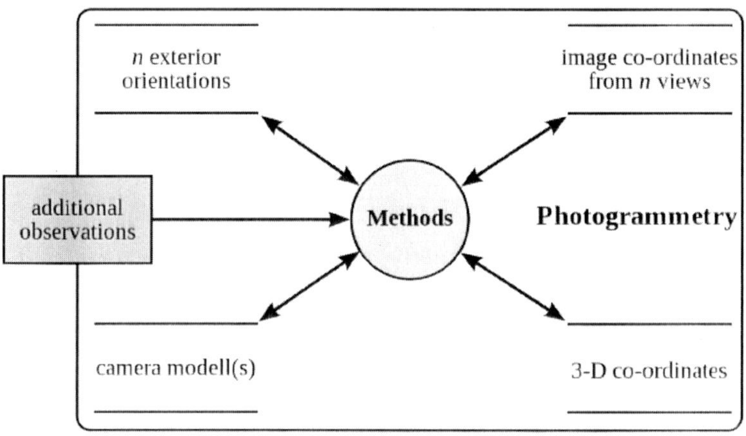

Georg Wiora's data model of photogrammetry

Photogrammetry uses methods from many disciplines, including optics and projective geometry. The data model on the right shows what type of information can go into and come out of photogrammetric methods.

The *3-D co-ordinates* define the locations of object points in the 3-D space. The *image co-ordinates* define the locations of the object points' images on the film or an electronic imaging device. The *exterior orientation* of a camera defines its location in space and its view direction. The *inner orientation* defines the geometric parameters of the imaging process. This is primarily the focal length of the lens, but can also include the description of lens distortions. Further *additional observations* play an important role: With *scale bars*, basically a known distance of two points in space, or known *fix points*, the connection to the basic measuring units is created. Each of the four main variables can be an *input* or an *output* of a photogrammetric method.

Photogrammetry has been defined by the American Society for Photogrammetry and Remote Sensing (ASPRS) as the art, science, and technology of obtaining reliable information about physical objects and the environment through processes of recording, measuring and interpreting photographic images and patterns of recorded radiant electromagnetic energy and other phenomena.

Applications

Photogrammetry is used in different fields, such as topographic mapping, architecture, engineering, manufacturing, quality control, police investigation, and geology, as well as by archaeologists to quickly produce plans of large or complex sites and by meteorologists as a way to determine the actual wind speed of a tornado, where objective weather data cannot be obtained. It is also used to combine live action with computer-generated imagery in movie post-production; The Matrix is a good example of the use of photogrammetry in film (details are given in the DVD extras).

This method is commonly employed in collision engineering, especially with automobiles. When litigation for accidents occurs and engineers need to determine the exact deformation present in the vehicle, it is common for several years to have passed and the only evidence that remains is accident scene photographs taken by the police. Photogrammetry is used to determine how much the car in question was deformed, which relates to the amount of energy required to produce that deformation. The energy can then be used to determine important information about the crash (such as the velocity at time of impact).

❑❑❑

Chapter 7

Image Analysis and Digital Image Processing

Image Analysis

Image analysis is concerned with making quantitative measurements from an image to produce a description of it. In the simplest form, this task could be reading a label on a grocery item, sorting different parts on an assembly line, or measuring the size and orientation of blood cells in a medical image. More advanced image analysis systems measure quantitative information and use it to make a sophisticated decision, such as controlling the arm of a robot to move an object after identifying it or navigating an aircraft with the aid of images acquired along its trajectory.

Image analysis techniques require extraction of certain features that aid in the identification of the object. Segmentation techniques are used to isolate the desired object from the scene so that measurements can be made on it subsequently. Quantitative measurements of object features allow classification and description of the image.

Image Segmentation

***Image segmentation* is the process that subdivides an image into its constituent parts or objects.** The level to which this subdivision is carried out, depends on the problem being solved, i.e., the segmentation should stop, when the objects of interest in an application have been isolated e.g., in autonomous air-to-ground target acquisition, suppose our interest lies in identifying vehicles on a road, the first step is to segment the road from the image and then to segment the contents of the road down to potential vehicles. Image thresholding techniques are used for image segmentation.

Classification

***Classification* is the labelling of a pixel or a group of pixels based on its grey value**. Classification is one of the most often used methods of information extraction. In Classification, usually multiple features are used for a set of pixels i.e., many images of a particular object are needed. In Remote Sensing area, this procedure assumes that the imagery of a specific geographic area is collected in multiple regions of the electromagnetic spectrum and that the images are in good registration. Most of the information extraction techniques rely on analysis of the spectral reflectance properties of such imagery and employ special algorithms designed to perform various types of 'spectral analysis'. The process of multi-spectral classification can be performed using either of the two methods: Supervised or Unsupervised.

In *Supervised classification*, the identity and location of some of the land cover types, such as urban, wetland, forest, etc., are known as priori through a combination of field works and toposheets. The analyst attempts to locate specific sites in the remotely sensed data that represents homogeneous examples of these land cover types. These areas are commonly referred as TRAINING SITES because the spectral characteristics of these known areas are used to 'train' the classification algorithm for eventual land cover mapping of reminder of the image. Multivariate statistical parameters are calculated for each training site. Every pixel both within and outside these training sites is then evaluated and assigned to a class of which it has the highest likelihood of being a member.

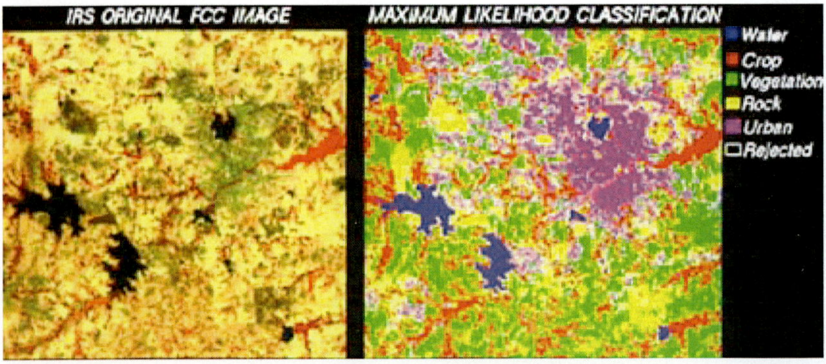

Fig. Image Classification

In an *Unsupervised classification*, the identities of land cover types has to be specified as classes within a scene are not generally known as priori because ground truth is lacking or surface features within the scene are not well-defined. The computer is required to group pixel data into different spectral classes according to some statistically determined criteria.

The comparison in medical area is the labelling of cells based on their shape, size, colour and texture, which act as features. This method is also useful for MRI images.

Image Restoration

Image restoration refers to removal or minimization of degradations in an image. This includes de-blurring of images degraded by the limitations of a sensor or its environment, noise filtering, and correction of geometric distortion or non-linearity due to sensors.

Image is restored to its original quality by inverting the physical degradation phenomenon, such as defocus, linear motion, atmospheric degradation and additive noise.

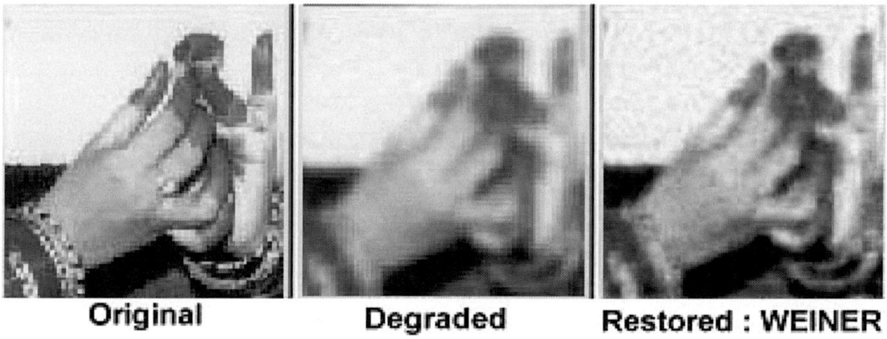

Fig. Weiner – Image Restoration

Image Reconstruction from Projections

Image reconstruction from projections [3] is a special class of image restoration problems, where a two- (or higher) dimensional object is reconstructed from several one-dimensional projections. Each projection is obtained by projecting a parallel X-ray (or other penetrating radiation) beam through the object. Planar projections are, thus, obtained by viewing the object from many different angles. Reconstruction algorithms derive an image of a thin axial slice of the object, giving an inside view otherwise unobtainable without performing extensive surgery. Such techniques are important in medical imaging (CT scanners), astronomy, radar imaging, geological exploration, and non-destructive testing of assemblies.

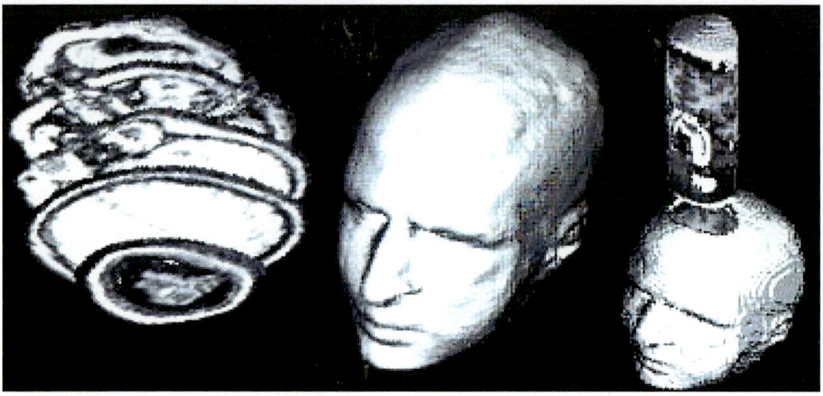

Fig. MRI Slices

Image Compression

Compression is a very essential tool for archiving image data, image data transfer on the network, etc. They are various techniques available for lossy and lossless compressions. One of most popular compression techniques, JPEG (Joint Photographic Experts Group) uses Discrete Cosine Transformation (DCT) based compression technique. Currently wavelet based compression techniques are used for higher compression ratios with minimal loss of data.

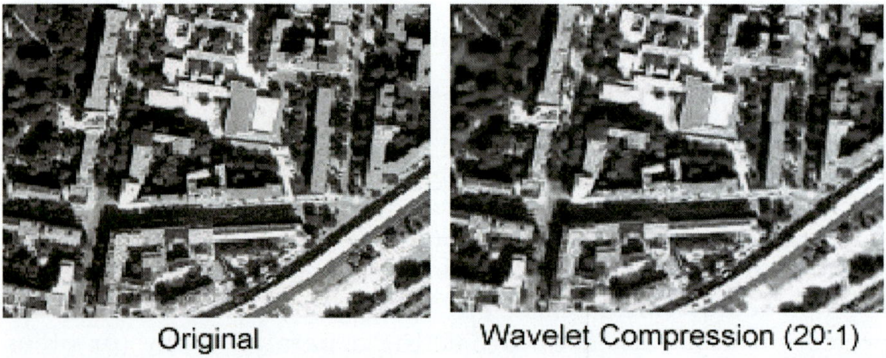

Fig. Wavelet Image Compression

Digital Image Processing (Dip)

Digital Image Processing

Digital image processing deals with manipulation of digital images through a digital computer. It is a subfield of signals and systems but focus particularly on images. DIP focuses on developing a computer system that is able to perform processing on an image. The input of that system is a digital image and the system process that image using efficient algorithms, and gives an image as an output. The most common example is Adobe Photoshop. It is one of the widely used applications for processing digital images.

How it works?

In the above figure, an image has been captured by a camera and has been sent to a digital system to remove all the other details, and just focus on the water drop by zooming it in such a way that the quality of the image remains the same.

Audience

This gives you the knowledge of widely used methods and procedures for interpreting digital images for image enhancement and restoration and performing operations on images, such as (blurring, zooming, sharpening, edge detection, etc.). It also focuses on the understanding of how the human vision works. How do human eye visualize so many things, and how do brain interpret those images? The tutorial also covers some of the important concepts of signals and systems, such as (Sampling, Quantization, Convolution, Frequency domain analysis, etc.).

Pre-requisites

Signals and systems

Since DIP is a subfield of signals and systems, so it would be good, if you already have some knowledge about signals and systems, but it is not necessary. But you must have some basic concepts of digital electronics.

Calculus and probability

Basic understanding of calculus, probability and differential equations is also required for better understanding.

Basic programming skills

Other than this, it requires some of the basic programming skills on any of the popular languages such as C++, Java , or MATLAB.

Flow chart of Digital Image Processing

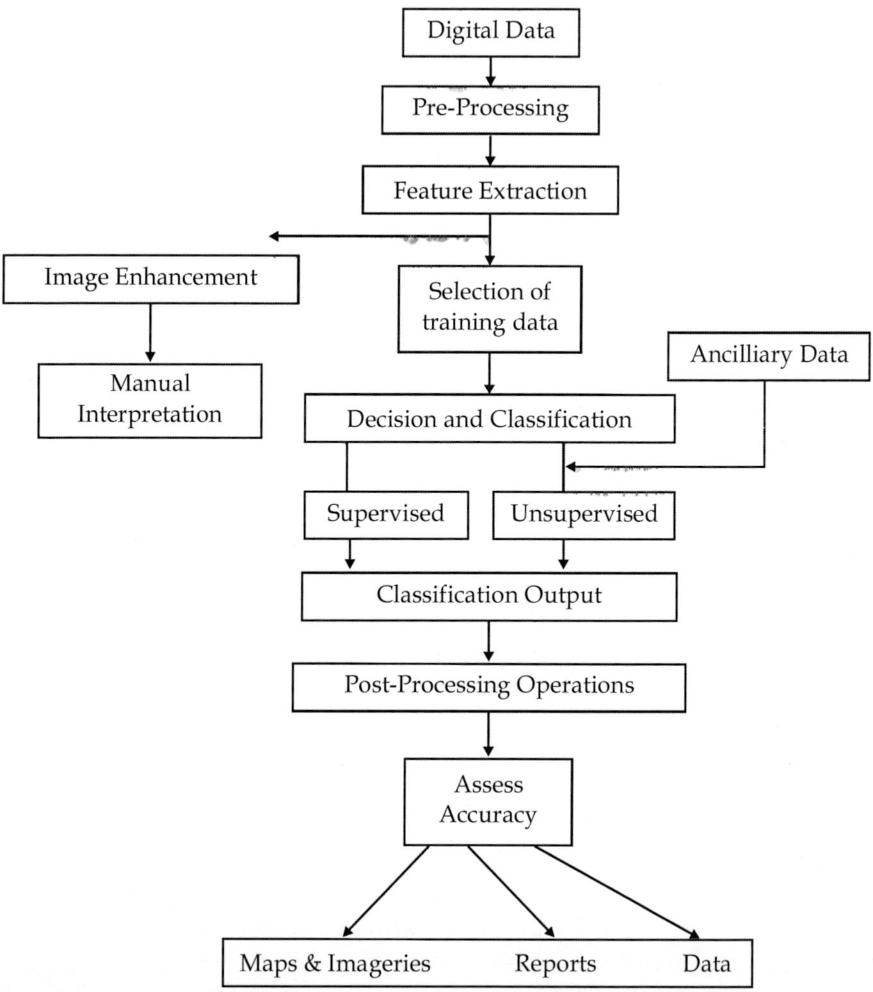

Digital Image Processing (DIP)

In today's world of advanced technology, where most remote sensing data are recorded in digital format, virtually all image interpretation and analysis involves some element of digital processing. Digital image processing may involve numerous procedures including formatting and correcting of the data, digital enhancement to facilitate better visual interpretation, or even automated classification of targets and features entirely by computer. In order to process remote sensing imagery digitally, the data must be recorded and available in a digital form suitable for storage on a computer tape or disk. Obviously, the other requirement for digital image processing is a computer system, sometimes referred to as an **image analysis system**, with the appropriate hardware and software to process the data. Several commercially available software systems have been developed specifically for remote sensing image processing and analysis.

For discussion purposes, most of the common image processing functions available in image analysis systems can be categorized into the following four categories:

1. Pre-processing
2. Image Registration
3. Image Enhancement
4. Image Transformation
5. Image Classification and Analysis

1. **Pre-processing : Pre-processing** functions involve those operations that are normally required prior to the main data analysis and extraction of information, and are generally grouped **as radiometric or geometric corrections**. Radiometric corrections include correcting the data for sensor irregularities and unwanted sensor or atmospheric noise, and converting the data, so they accurately represent the reflected or emitted radiation measured by the sensor. Geometric corrections include correcting for geometric distortions due to sensor-Earth geometry variations, and conversion of the data to real world coordinates (e.g. latitude and longitude) on the Earth's surface.

2. **Image Registration :** Step of transferring latitude and longitude from topo map by map image rectification/registration.
3. **Image Enhancement :** The objective of the second group of image processing functions grouped under the term of **image enhancement**, is solely to **improve the appearance of the imagery** to assist in visual interpretation and analysis. Examples of enhancement functions include contrast stretching to increase the tonal distinction between various features in a scene, and **spatial filtering** to enhance (or suppress) specific spatial patterns in an image.
4. **Image transformations : Techniques applied using more than one band at a time in order to extract feature of our interest. Techniques of Arithmetic transformations, band ratio of red band and near IR, NDVI (Normalized Difference Vegetation Index) and PCA (Principle Component Analysis) are used.**

   e.g. NDVI = (NIR-R) / (NIR+R), this ratio is very high for healthy crop.

Image transformations are operations similar in concept to those for image enhancement. However, unlike image enhancement operations which are normally applied only to a single channel of data at a time, image transformations usually involve combined processing of data from multiple spectral bands. Arithmetic operations (i.e. subtraction, addition, multiplication, division) are performed to combine and transform the original bands into "new" images which better display or highlight certain features in the scene. We will look at some of these operations including various methods of **spectral or band** rationing, and a procedure called **principal components analysis,** which is used to more efficiently represent the information in multi-channel imagery.

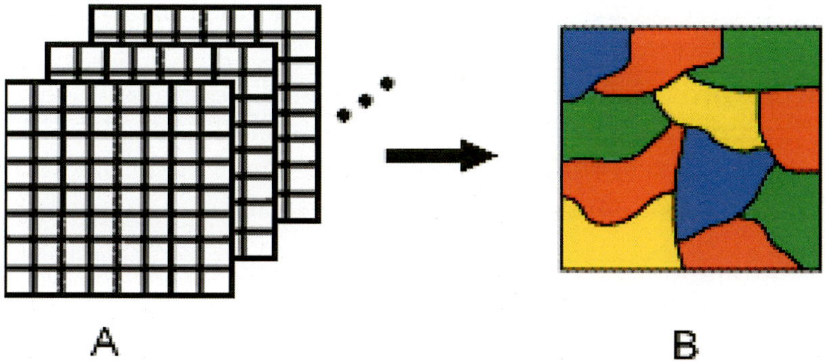

5. **Image classification and analysis : Image classification and analysis** operations are used to digitally identify and classify pixels in the data. **Classification** is usually performed on multi-channel data sets (A) and this process assigns each pixel in an image to a particular class or theme (B) based on statistical characteristics of the pixel brightness values. There are a variety of approaches taken to perform digital classification. We will briefly describe the two generic approaches which are used most often, namely **supervised** and **unsupervised** classification.

Introduction to Image Interpretation

Image Interpretation

Image interpretation is defined as the act of examining images to identify objects and judge their significance. An interpreter studies remotely sensed data and attempts through logical process to detect, identify, measure and evaluate the significance of environmental and cultural objects, patterns and spatial relationships. It is an information extraction process.

Aerial photographs, as well as imagery, obtained by remote sensing employing electromagnetic energy as the means of detecting and measuring target/objects characteristics, has applicability to various fields because of four basic reasons :

First It represents a larger area of the earth from a perspective view and provides a format that facilitates the study of objects and their relationships.

Second Certain types of imagery and aerial photograph can provide a 3-D view.

Third Characteristics of objects not visible to the human eye can be transformed into images

Fourth It provides the observer with a permanent record/ representation of objects at any moment of time. In addition, data is real-time, repetitive and, when in digital form, is computer compatible for quick analysis.

Basic principles of image interpretation

Images and their interpretability

- An image taken from the air or space is a pictorial presentation of the pattern of a landscape.
- The pattern is composed of indicators of objects and events that relate to the physical, biological and cultural components of the landscape.
- Similar conditions, in similar circumstances and surroundings, reflect similar patterns, and unlike conditions reflect unlike patterns.
- The type and amount of information that can be extracted is proportional to the knowledge, skill and experience of the analyst, the methods used for interpretation and the analyst's awareness of any limitations.

Factors governing the Quality of an image

In addition to the inherent characteristics of an object itself, the following factors influence image quality:

- Sensor characteristics (film types, digital systems)
- Season of the year and time of day
- Atmospheric effects
- Resolution of the imaging system and scale
- Image motion
- Stereoscopic parallax

Factors governing Interpretability

1. Visual and mental acquity of the interpreter
2. Equipment and technique of interpretation
3. Interpretation keys, guides, manuals and other aids.

Visibility of Objects

The objects on aerial photographs or imagery are represented in the form of photo images in tones of grey in B/W photography and in colour/false colour photography in different colours/hues. This visibility of objects in the images varies due to :

a) The inherent characteristics of the objects

b) The quality of the aerial photography or imagery

Inherent characteristics of objects

In any photographic image forming process, the negative is composed of tiny silver deposits formed by the action of light on photosensitive film during exposure. The amount of light received by the various sections of the film depends on the reflection of electromagnetic radiation (EMR) from various objects. This light, after passing through the optical system, gives rise to different tones and textures.

Quality of Aerial Photography/Imagery

The quality of image interpretation depends on the quality of the basic material on which the images are formed. Normally, in visual interpretation, these images are formed on the photograph and represented in tones of grey or in colours of various hues, chroma and values. A study of the factors, affecting image quality and characteristics of images, is essential from an interpreter's point of view.

Tonal or Colour Contrast between an Image and its background

Photographic tone contrast is the difference in brightness between an image and its background. Similarly, in colour photography colour contrast is the result of all hue values and chroma differences between the image and its background. The tonal contrast can be sufficiently increased with proper filters.

Image Sharpness Characteristics

Sharpness is the abruptness with which tone or colour contrasts appear on the photograph or imagery. Both tone and sharpness enable an interpreter to distinguish one object from another. To a large extent, image sharpness is dependent on the focusing ability of the optical system. Image sharpness is closely related to the resolution of the optical system.

Stereoscopic Parallax Characteristics

Stereoscopic parallax is the displacement of the apparent position of an image with respect to a reference point of observation. Sufficient parallax is necessary in order distinguish objects from their shadows. Parallax depends on the height of an object, lying height and the stereobase or its corollary, the forward overlap. Stereoscopic parallax can be improved by choosing the right base/height (B/H) ratio. The above investigation appears to be over simplified as a number of other factors can be mentioned which obviously effect the image quality. However, for the purpose of simplification, we may conclude that other factors influence image quality indirectly through their effect on tone, sharpness or parallax.

In general, if image motion and exposure times were no problem, we would obviously use fine grain, high definition, slow photographic material, with an appropriate filter in order to get better sharpness and contrast.

Elements of Image Interpretation

The word 'photograph' in Greek means to draw with light, and a photograph, in fact, is nothing more or less than a graphic record of energy intensities. An image represents energy reflected, emitted or transmitted from an object in different parts of the spectrum.

Image interpretation is essential for the efficient and effective use of the data. While the above properties of aerial photographs/imagery help an interpreter to detect objects due to their tonal variations, he must also take advantage of other important characteristics of the objects in order to recognize them.

The following nine elements of image interpretation are regarded as being of general significance, irrespective of the precise nature of the imagery and the features it portrays :

1. **Shape:** Numerous components of the environment can be identified with reasonable certainty merely by their shape. This is true of both natural features and man-made objects.

2. **Size :** In many cases, the length, breadth, height, area and/or volume of an object can significant, whether these are surface features (e.g. different tree species) or atmospheric phenomena (e.g. cumulus versus cumulonimbus clouds). The approximate size of objects can be judged by comparisons with familiar features (e.g. roads) in the scene.
3. **Tone :** We have seen how different objects emit or reflect different wavelengths intensities of radiant energy. Such differences may be recorded as variations of tone, colour or density, which enable discrimination of many spatial variables, for example, on land different crop types or at sea water bodies of contrasting depths or temperatures. The terms 'light', 'medium' or 'dark' are used to describe variations in tone.
4. **Shadow :** Hidden profiles may be revealed in silhouette (e.g. the shapes of buildings or the forms of field boundaries). Shadows are especially useful in geomorphological studies, where micro relief features may be easier to detect under conditions of low-angle solar illumination than when the sun is high in the sky. Unfortunately, deep shadows in areas of complex detail may obscure significant features, e.g. the volume and distribution of traffic on a city Street.
5. **Pattern :** Repetitive patterns of both natural and cultural features are quite common, which is fortunate because much image interpretation is aimed at the mapping and analysis of relatively complex features rather than the more basic units of which they may be composed. Such features include agricultural complexes (e.g. farms and orchards) and terrain features (e.g. alluvial river valleys and coastal plains).
6. **Texture :** Texture is an important image characteristic closely associated with tone in the sense that it is a quality that permits two areas of the same overall tone to be differentiated on the basis of microtonal patterns. Common image textures include smooth, rippled, mottled, lineated and irregular. Unfortunately, texture analysis tends to be rather subjective, since different interpreters may use the same terms in slightly different ways. Texture is rarely the only criterion of identification or correlation employed in interpretation. More often, it is invoked as the basis for a subdivision of categories already established using more fundamental criteria. For example, two rock units may have the same tone but different textures.

7. **Site:** At an advanced stage in image interpretation, the location of an object with respect to terrain features of other objects may be helpful in refining the identification and classification of certain picture contents. For example, some tree species are found more commonly in one topographic situation than in others, while in industrial areas the association of several clustered, identifiable structures may help us determine the precise nature of the local enterprise. For example, the combination of one or two tall chimneys. large central building, conveyors, cooling towers and solid fuel piles point to the correct identification of a thermal power station.
8. **Resolution :** Resolution of a sensor system may be defined as its capability to discriminate two cosely spaced objects from each other. More than most other picture characteristics, resolution depends on aspects of the remote sensing system itself, including its nature, design and performance, as well as the ambient conditions during the sensing programme and subsequent processing of the acquired data. An interpreter must have a knowledge about the resolution of various remote sensing data products.
9. **Stereoscopic Appearance :** When the same feature is photographed from two different positions with overlap between successive images, an apparently solid model of the feature can be seen under a stereoscope. Such a model is termed as stereo model and the three-dimensional view it provides, can aid interpretation. This valuable information cannot be obtained from a single print.

Techniques of image interpretation

Image interpretation comprises at least three mental acts that may or may not be performed simultaneously :

i) The measurement of images of objects
ii) Identification of the objects imaged
iii) Appropriate use of this information in the solution of the problem.

In visual interpretation, the methodology of interpretation for each separate discipline will depend on :

- Kind of information to he interpreted
- Accuracy of the results to he obtained

- The reference level of the person executing the interpretation
- Kind and type of imagery or photographs available Instruments available
- Instruments available
- Scale and other requirements of the final map
- External knowledge available and any other sensory surveys that have been or wil1 be made in the near future in the same area.

Activities of image interpretation

Image-interpretation is a complex process comprising physical as well as mental activities. This means familiarity with so wide a variety of stimuli that the even most accomplished interpreter is occasionally dependent on reference materials.

The reference material in the form of identification keys is a useful aid in image Interpretation. Many types of image interpretation keys are available or may be constructed depending on the abilities of the interpreter and the purpose to be served by the interpretation.

Manner of organization or presentation of image interpretation keys:

All image interpretation keys are based upon diagnostic features of the images of objects or conditions to be identified :

1. Selective Keys
2. Elimination Keys

Methods of search and sequence of interpretation

Sequence of Activities

Normally the activities in an image-interpretation sequence include the following:

Detection

Detection means selectively picking out an object or element of importance for the particular kind of interpretation in hand. It is often coupled with recognition, in which case the object is not only seen but also recognized.

Recognition and Identification

Recognition and identification together are sometimes termed as photo-reading. However, they are fundamentally the same process and refer to the process of classification of an object by means of specific or local knowledge within a known category upon an object's detection in a photo-image.

Analysis

Analysis is the process of separating or delineating a set of similar objects. In analysis, boundary lines are drawn separating the groups, and the degree of reliability of these lines may be indicated.

Deduction

Deduction may be directed to the separation of different groups of objects or elements and the deduction of their significance based on converging evidence. The evidence is derived mainly from visible objects or from invisible elements, which give only partial information on the nature of certain correlative indications.

Classification

Classification establishes the identity of a surface or an object delineated by analysis. It includes modification of the surface into a pertinent system for use in field investigation. Classification is made in order to group surfaces or objects according to those aspects that, for a certain point of view, bring out their most characteristic aspects.

Idealization

Idealization refers to the process of drawing or standardized representations of what is actually seen in the photo image. This process is helpful for the subsequent use of photograph/imagery during field investigations and in the preparation of base maps.

Convergence of evidence

The principle of convergence of evidence requires the interpreter first to recognize basic features or types of features and then to consider their arrangement (pattern) in the real context. Several interpretations may suggest themselves. Critical examination of the evidence usually shows that all interpretations but one are unlikely or impossible. The greatest difficulty in interpreting images involves judging degrees of probability.

Sensors in Photographic Image Interpretation

As stated earlier, characteristics not visible to the human eye can also be recorded and displayed by using proper sensor types. Digital data can also be transferred onto any type of film, depending on the type of study to be carried out. Normally, the four types of films are used for visual data display as follows :

a) Black-and-white panchromatic

b) Black-and-white infrared

c) Colour

d) Colour infrared/false colour

Methods of analysis and reference levels

The mental process involved in image-interpretation is related to the reference levels of the interpreter, i.e. the capacity of the interpreter to make decisions, consciously or unconsciously. The reference level in context is the amount of knowledge stored in the mind of any group of personnel involved in the interpretation of photographs or imagery. Methods of analysis depend on the type of material used for interpretation as well as the instruments and equipment available. The following methods are generally used :

***Monocular analysis*:** In the case of satellite imagery and photographic enlargements.

***Stereoscopic analysis*:** For vertical/near vertical aerial photographs, SPOT as well as IRS-IC stereo-imagery.

***Densitometric analysis*:** For both aerial photographs and imagery for quantitative analysis using densitometers, based on the measurements of tonal variation, a fundamental characteristic of all objects, for identifying terrain features. Such analysis is now carried out using digital image processing systems.

Instruments for visual interpretation and transfer of data

Interprétation Instruments

Monocular instruments : Magnifiers

Stereoscopic instruments : Mirror & pocket stereoscope Interpretoscope

Zoom stereoscope

Scanning mirror

Stereoscope

Instruments for Transfer of Data

For flat terrain : Sketchmaster Stereosketchmaster

Zoom transferscope

Optical pantograph or reflecting projector

For hilly terrain : Stereoplotters

Orthophoto together with its stereo-mate, can be used for interpretation and delineation's. Since preparation of orthophoto and its stereo-mate is a complex process, the method is not so popular.

□□□

Chapter 8

Agencies and Institutes Involved in Remote Sensing at National and International Level

Indian Space Research Organization (ISRO)

ISRO is managed by the Department of Space (DoS) of the Government of India. DoS itself falls under the authority of the Prime Minister and the Space Commission, and manages the following agencies and institutes:

- Indian Space Research Organisation
  - Vikram Sarabhai Space Centre (**VSSC**), Thiruvananthapuram.
  - Liquid Propulsion Systems Centre (**LPSC**), Thiruvananthapuram.
  - Satish Dhawan Space Centre (**SDSC-SHAR**), Sriharikota.
  - ISRO Propulsion Complex (**IPRC**), Mahendragiri.
  - ISRO Satellite Centre (**ISAC**), Bengaluru.
  - Space Applications Centre (**SAC**), Ahmedabad.
  - National Remote Sensing Centre (**NRSC**), Hyderabad.
  - ISRO Inertial Systems Unit (**IISU**), Thiruvananthapuram.
  - Development and Educational Communication Unit (**DECU**), Ahmedabad.
  - Master Control Facility (**MCF**), Hassan.
  - ISRO Telemetry, Tracking and Command Network (**ISTRAC**), Bengaluru.
  - Laboratory for Electro-Optics Systems (**LEOS**), Bengaluru.
  - Indian Institute of Remote Sensing (**IIRS**), Dehradun.

- Antrix Corporation - The marketing arm of ISRO.
- Physical Research Laboratory (PRL), Ahmedabad.
- National Atmospheric Research Laboratory (NARL), Gadanki.
- North-Eastern Space Applications Centre[18] (NE-SAC), Umiam.
- Semi-Conductor Laboratory (SCL), Mohali.
- Indian Institute of Space Science and Technology (IIST), Thiruvananthapuram - India's space university.

Research Facilities

Facility	Location	Description
Vikram Sarabhai Space Centre	Thiruvananthapuram	The largest ISRO base is also the main technical centre and the venue of development of the SLV-3, ASLV, and PSLV series. The base supports India's Thumba Equatorial Rocket Launching Station and the Rohini Sounding Rocket programme. This facility is also developing the GSLV series.
Liquid Propulsion Systems Centre	Thiruvananthapuram and Bengaluru	The LPSC handles design, development, testing and implementation of liquid propulsion control packages, liquid stages and liquid engines for launch vehicles and satellites. The testing of these systems is largely conducted at IPRC at Mahendragiri. The LPSC, Begaluru also produces precision transducers.
Physical Research Laboratory	Ahmedabad	Solar planetary physics, infrared astronomy, geo-cosmo physics, plasma physics, astrophysics, archaeology, and hydrology are some of the branches of study at this institute. An observatory at Udaipur also falls under the control of this institution.
Semi-Conductor Laboratory	Chandigarh	Research & Development in the field of semiconductor technology, micro-electromechanical systems and process technologies relating to semiconductor processing.
National Atmospheric Research Laboratory	Chittoor	The NARL carries out fundamental and applied research in Atmospheric and Space Sciences.
Space Applications Centre	Ahmedabad	The SAC deals with the various aspects of practical use of space technology.[19] Among the fields of research at the SAC are geodesy, satellite based telecommunications, surveying, remote sensing, meteorology, environment monitoring, etc. The SEC additionally operates the Delhi Earth Station.
North-Eastern Space Applications Centre	Shillong	Providing developmental support to North East by undertaking specific application projects using r emote sensing, GIS, satellite communication and conducting space science research.

Test Facilities

Facility	Location	Description
ISRO Propulsion Complex	Mahendragiri	Formerly called LPSC -Mahendragiri, was declared a separate centre. It handles testing and assembly of liquid propulsion control packages, liquid engines and stages for launch vehicles and satellites.

Construction and Launch Facilities

Facility	Location	Description
ISRO Satellite Centre	Bengaluru	The venue of eight successful spacecraft projects is also one of the main satellite technology bases of ISRO. The facility serves as a venue for implementing indigenous spacecraft in India. The satellites Ayrabhata, Bhaskara, APPLE, and IRS-1A were constructed at this site, and the IRS and INSAT satellite series are presently under development here.
Laboratory for Electro-Optics Systems	Bengaluru	The Unit of ISRO responsible for the development of altitude sensors for all satellites. The high precision optics for all cameras and payloads in all ISRO satellites including Chandrayaan-1 are developed at this laboratory. Located at Peenya Industrial Estate, Bengaluru.
Satish Dhawan Space Centre	Sriharikota	With multiple sub -sites the Sriharikota isl and facility acts as a launching site for India's satellites. The Sriharikota facility is also the main launch base for India's sounding rockets. The centre is also home to India's largest Solid Propellant Space Booster Plant (SPROB) and houses the Static Test and Evaluation Complex (STEX).
Thumba Equatorial Rocket Launching Station	Thiruvananthapuram	TERLS is used to launch sounding rockets.

Tracking and Control Facilities

Facility	Location	Description
Indian Deep Space Network (IDSN)	Bengaluru	This network receives, processes, archives and distributes the spacecraft health data and payload data in real time. It can track and monitor satellites up to very large distances, even beyond the Moon.
National Remote Sensing Centre	Hyderabad	The NRSC applies remote sensing to managenatural resources and study aerial surveying. [19] With centres at Balanagar and Shadnagar it also has training facilities at Dehradun in form of the Indian Institute of Remote Sensing.
ISRO Telemetry, Tracking and Command Network	Bengaluru (headquarters) and a number of ground stations throughout India and World.	Software development, ground operations, Tracking Telemetry and Command (TTC), and support is provided by this institution. ISTRAC has Tracking stations throughout the country and all over the world in Port Louis (Mauritius), Bearslake (Russia), Biak (Indonesia) and Brunei.
Master Control Facility	Bhopal; Hassan	Geostationary satellite orbit raising, payload testing, and in-orbit operations are performed at this facility. [22] The MCF has earth stations and Satellite Control Centre (SCC) for controlling satellites. A second MCF-like facility named 'MCF-B' is being constructed at Bhopal.

Human Resource Development (HRD)

Facility	Location	Description
Indian Institute of Remote Sensing (IIRS)	Dehradun	Indian Institute of Remote Sensing (IIRS), an independent unit of Indian Space Research Organisation (ISRO), Department of Space, Govt. of India is a premier training and educational institute set up for developing trained professionals (P.G and Ph.D level) in the field of Remote Sensing, Geoinformatics and GPS Technology for Natural Resources, Environmental and Disaster Management. IIRS is also executing many R&D projects on Remote Sensing and GIS for societal applications.
Indian Institute of Space Science and Technology (IIST)	Thiruvananthapuram	The institute offers undergraduate and graduate courses in Aerospace engineering, Avionics and Physical Sciences. The students of the first three batches of IIST have been inducted into different ISRO centres as of September 2012.
Development and Educational Communication Unit	Ahmedabad	The centre works for education, research, and training, mainly in conjunction with the INSAT programme. [19] The main activities carried out at DECU include GRAMSAT and EDUSAT projects. [20] The Training and Development Communication Channel (TDCC) also falls under the operational control of the DECU.

Commercial Wing (Antrix Corporation)

Facility	Location	Description
Antrix Corporation	Bengaluru	The marketing agency under government control markets ISRO's hardware, manpower, software, launch services, etc.

Other Facilities

- Balasore Rocket Launching Station (BRLS) - Odisha
- ISRO Inertial Systems Unit (IISU) - Thiruvananthapuram
- Indian Regional Navigational Satellite System (IRNSS)
- Aerospace Command of India (ACI)
- Indian National Committee for Space Research (INCOSPAR)
- Inter University Centre for Astronomy and Astrophysics (IUCAA)
- Indian Space Science Data Centre (ISSDC)
- Spacecraft Control Centre (SCC)
- Regional Remote Sensing Service Centres (RRSSC)
- Development and Educational Communication Unit (DECU)

S.No.	Satellite	Date of Launch	Launch Vehicle	Status
1	IRS 1A	17 March, 1988	Vostok, USSR	Mission Completed
2	IRS 1B	29 August, 1991	Vostok, USSR	Mission Completed
3	IRS P1 (also IE)	20 September, 1993	PSLV-D1	Crashed, due to launch failure of PSLV
4	IRS-P2	15 October, 1994	PSLV-D2	Mission Completed
5	IRS-1C	28 December, 1995	Molniya, Russia	Mission Completed
6	IRS-P3	21 March, 1996	PSLV-D3	Mission Completed
7	IRS 1D	29 September, 1997	PSLV-C1	Mission Completed
8.	IRS-P4 (Oceansat-1)	27 May, 1999	PSLV-C2	Mission Completed
9.	Technology Experiment Satellite (TES)	22 October, 2001	PSLV-C3	Mission Completed
10.	IRS P6 (Resourcesat-1)	17 October, 2003	PSLV-C5	In Service
11.	IRS P5 (Cartosat 1)	5 May, 2005	PSLV-C6	In Service
12.	Cartosat 2 (IRS P7)	10 January, 2007	PSLV-C7	In Service
13.	Cartosat 2A	28 April, 2008	PSLV-C9	In Service
14.	IMS 1	28 April, 2008	PSLV-C9	In Service
15.	Oceansat-2	23 September, 2009	PSLV-C14	In Service
16.	Cartosat-2B	12 July, 2010	PSLV-C15	In Service
17.	Resourcesat-2	20 April, 2011	PSLV-C16	In Service
18.	Megha-Tropiques	12 October, 2011	PSLV-C18	In Service
19.	RISAT-1	26 April, 2012	PSLV-C19	In Service
20.	SARAL	25 Feb, 2013	PSLV-C20	In Service

❑❑❑

Chapter 9

Remote Sensing, GIS, GPS and Aerial Photographs and their Role in Forest Management

Global Positioning System (GPS)

The Global Positioning System (GPS) is a satellite-based navigation system made up of a network of 27 satellites placed into orbit by the U.S. Department of Defense. GPS is primarily satellite based radio navigation system for real time positioning. It provides three dimensional positions, navigation and time in information to suitability equipped users. GPS was originally intended for military applications, but in the 1980s, the government made the system available for civilian use. GPS works in any weather conditions, anywhere in the world, 24 hours a day. There are no subscription fees or setup charges to use GPS.

How it works?

GPS satellites circle the earth twice a day in a very precise orbit and transmit signal information to earth. GPS receivers take this information and use triangulation to calculate the user's exact location. Essentially, the GPS receiver compares the time a signal was transmitted by a satellite with the time it was received. The time difference tells the GPS receiver how far away the satellite is. Now, with distance measurements from a few more satellites, the receiver can determine the user's position and display it on the unit's electronic map.

A GPS receiver must be locked on to the signal of at least three satellites to calculate a 2D position (latitude and longitude) and track movement. With four or more satellites in view, the receiver can determine the user's 3D position (latitude, longitude and altitude). Once the user's position has been determined, the GPS unit can calculate other information, such as speed, bearing, track, trip distance, distance to destination, sunrise and sunset time and more.

Disadvantages of conventional survey

1. Inter-visibility dependent
2. Weather dependent
3. Time consuming
4. Very laborious
5. Require trained observers
6. Require large man power

Functions of GPS

- For connecting & storing points.
- For collecting & storing routes, a path between two or more points.
- 'GOTO' function is useful in guiding to a pre-determined point.
- GPS can be programmed to 'beep' when you are within a certain distance of the defined way point.
- Marine GPS with built-in tide tables provide information & ranges for any date & any place.
- GPS can calculate ground speed as you walk, run, drive or fly.
- In addition to providing with your latitude & longitude, GPS provides you with altitude information.

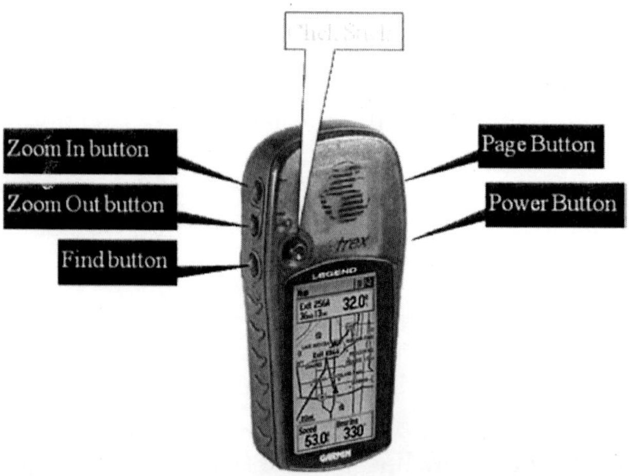

Basic concept of GPS

Velocity x Time = Distance

- Radio waves travel at the speed of light, roughly 3×10^8 m/sec.
- If it took 0.06 seconds to receive a signal transmitted by a satellite floating directly overhead, use this formula to find your distance from the satellite.

Therefore Velocity x Time = Distance

3×10^8 m/sec x 0.06 seconds = 18,000 kms

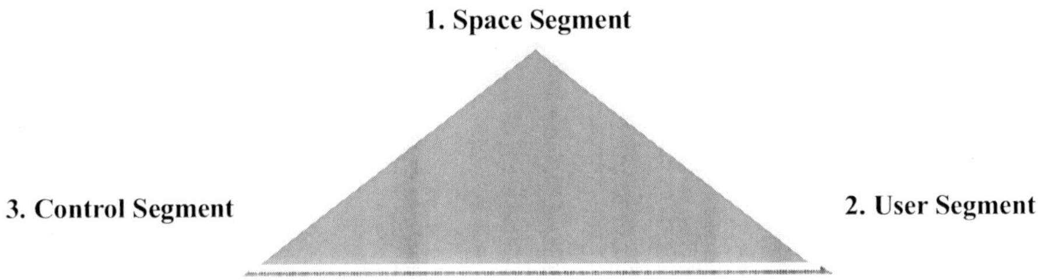

Applications of GPS in Forestry

1. In boundary survey and demarcation
2. Point determination in survey
3. Precise positioning is possible in remote and inaccessible areas
4. Gives accurate real time positions of latitude, longitude, altitude
5. Accurate mapping
6. Path finding
7. Sample plot surveys
8. Terrain details
9. Location of plantation sites
10. Regeneration plots
11. Insect disease tracing
12. Forest fires monitoring
13. Forest road alignment and length measurements
14. Research plots

Remote Sensing (RS)

Remote sensing is the process of obtaining measurements of an object from a distance.

It is a science of collecting and interpreting information about a target without being in physical contact with the object.

"Remote sensing is the science and art of obtaining information about an object, area, or phenomenon through the analysis of data acquired by a device that is not in contact with the object, area, or phenomenon under investigation".

Remote Sensing refers to gathering and processing of information about earth's environment and it's Natural & Cultural Resources through Aerial photography and Satellite scanning.

Stages in remote sensing

- Requirement of an energy source (The Sun is the main source of energy during day time.)
- Energy interaction with the atmosphere
- Interaction with the target
- Recording of energy by sensor
- Data transmission & processing
- Image processing & analysis
- Application

Advantages of the remote sensing

- **Synoptic view :** It facilitates the study of various features of the earth surface in their spatial relation to each other & helps to delineate the required features & phenomenon.
- **Accessibility :** It makes it possible to gather information about inaccessible areas, where it is not possible to gather information through ground surveys.
- **Time** : These techniques save time & efforts as information about large area can be gathered quickly.
- **Multidisciplinary applications** : Remote sensing data are useful to different disciplines, such as geology, fisheries, forestry, land use, etc.

Geographic Information System (GIS)

GIS is computer based integrated data base management system, in which large volumes of geo-referenced spatial data derived from variety of sources are, efficiently ***stored, organized, manipulated, retrieved, analyzed and displayed/presented*** according to the user defined specifications.

OR

GIS is also defined as ***"set of powerful tools for collecting, storing, retrieving at will, transforming and displaying spatial data from the real world, for a set of purposes".***

Components of GIS

GIS runs on the whole spectrum of computer hardware ranging from portable personal computers to a multi-user supercomputer & is programmed in a wide variety of software languages. There are number of elements that are essential for effective GIS operation.

- The presence of a processor with sufficient power to run the software.
- Sufficient memory for storage of large volume of data.
- A good quality high resolution colour screen.
- Data input & output devices.
- Live ware.

Advantages of GIS

1. Capacity to **handle large volume** of data in a sufficient manner
2. **Accuracy** and **consistency** in information gathering
3. **Speed** in analyze, retrieval and updating data
4. **Flexibility** in handling various types of information
5. **Ability** to simulate complex modules
6. **Compatibility** with the state of the art technology
7. **Cost effectiveness** in decision-making
8. It will be **transparent,** as it will be developed through logical algorithms of computers without much personal bias.

Components of GIS

- An operational GIS also has a series of components that combine to make the system work. These components are critical to a successful GIS.\
- A working GIS integrates five key components
- Hardware, Software, Data, People, Methods

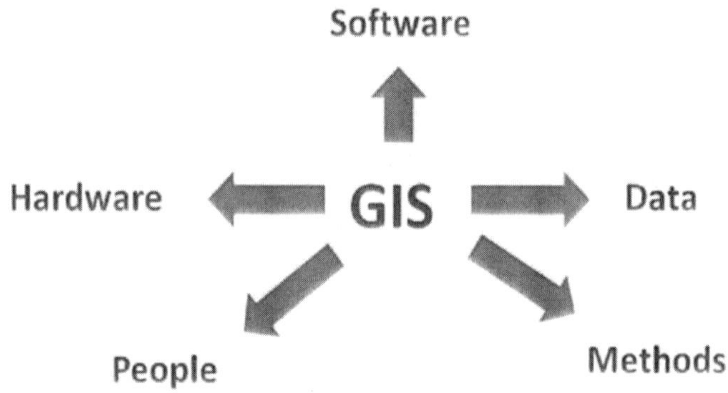

GIS subsystems

A GIS has four main functional subsystems. These are :

1. A data input subsystem
2. A data storage and retrieval subsystem
3. A data manipulation and analysis subsystem
4. A data output and display subsystem

Applications of GIS and Remote Sensing (RS) in forestry

1. Vegetation mapping

- **a. Extent of forest cover**
- **b. Forest stock mapping**
- **c. Growing stock estimation**
- **d. Habitat management**
- **e. Forest types**
- **f. Forest density**
- **g. Plantations**
- **j. Grass lands**
- **j. Protected areas**

2. Change monitoring & damage assessment

- **a. Afforestation & social forestry**
  - ***Catchment area treatment plans***
  - ***Wasteland development planning***
  - ***Wasteland reclamation***
  - ***Monitoring SF nurseries***
- **b. Forest encroachment**
- **c. Forest fire**
- **d. Shifting cultivation**
- **e. Forest cover transformation**
- **f. Forest environment impact assessment**

3. Wildlife management

- **a. Habitat assessment**
- **b. Habitat suitability**
- **c. Corridor identification**
- **d. Eco-development/eco-restoration**
- **e. Tiger census studies**

4. Management plans

- **a. Working plan**
- **b. JFPM & Micro level**
- **c. Land evaluation**
- **d. Eco-development plan**

5. Biodiversity studies

- **a. Rich & threat areas**
- **b. Biomass estimation**
- **c. Disturbance/biotic pressure**
- **d. NPP**
- **e. Mangrove forest preservation projects**
- **f. Preparation of biodiversity conservation plans for sanctuaries & national parks**

Aerial Photography

It is the method of remote sensing in which cameras are other devices, fixed in a aircraft flying at fixed altitude are used to take photographs of any resource on the earth.

Films are processed after landing. The interpretation of photographs gives information about the resource. As photographs are generally taken in this technique, hence, it is also called as aerial photographs. Word photography literally means **'Writing with light'**.

Aerial photography means photographing or taking pictures of the earth from the air.

Aerial photography is synonymously called as *Aerial remote Sensing*.

Classification of aerial photographs

1. Based on direction of exposure

- Vertical photographs
- Oblique photographs

2. Number and type of camera and lenses used

- Tri camera photography
- Convergent photography
- Multiple lens photography
- Continuous strip photography
- Small camera photography

3. Field of view of the camera

- Normal angle photography
- Wide angle photography
- Super angle photography

4. Based on the types of sensors and film emulsions

- Panchromatic photography
- Infrared photography
- Colour photography

- False colour photography
- Multi band photography

5. Scale of photography

- Very large photography
- Large photography
- Medium photography
- Small photography
- Very small photography

6. Season of photography

- Autumn photography
- Winter photography
- Spring photography
- Summer photography
- Rain photography

7. According to photographic paper

- Single weight photography
- Double weight photography

8. According to security

- Top secret photography
- Secret photography
- Restricted photography
- De restricted photography

Role/Use of Aerial Photographs in Forestry Related Activities and Management of Forests

Use of aerial photography in forest inventory

Aerial photographs together with knowledge of photo interpretation and photogrammetry have provided a new and very useful tool for preparation of forest inventory. In order to prepare a detailed and useful inventory of forest resources in a country, the following information is required :

a) Location of forests, their distribution and area

For this purpose a map of forests in general and functional classification in particular is required. In the productive forests, maps of forest areas by forest types, by productivity of sites and by maturity classes *etc* are necessary to assess the yield of forests.

b) The growing stock in the production of forests

For this purpose, enumeration or inventory of trees by species and by diameter classes is required. This would facilitate estimation of their volume by locations so that management plans could be drawn or industrial plan may be prepared.

c) Information about various factors affecting volume production

For this, information on site, diameter and height, increment, age, timber quality and also about areas where new forests could be raised is necessary.

Thus, the first prerequisite of a comprehensive and useful forest inventory is map of forest area and estimation of area covered by the forests. In this context, aerial photography helps in preparation of maps of forest areas and inventory of forests.

Use of aerial photographs in forest management

Forest management is that branch of forestry which deals with the organization of forest property for its proper maintenance by prescribing in time and place, the various operations are necessary for conservation, protection and improvement of forest on the one hand and controlled harvesting of the forest on the other. The structure of management of forest is erected on the basis of maps and its inventory. Once the maps are prepared and inventory of forest resources compiled, a management plan can be prepared by applying scientific, technical and economic principles of forestry.

Aerial photographs play an important role in the following forestry activities :

i) Organization and assessment of plantation work

Aerial photographs play an important role in plantations. First of all, they provide a permanent record of the type and condition of the forest crop converted into plantations for comparison of results. With the help of aerial photographs, steep and broken terrain, rocky outcrops, water-logged areas and natural blanks can be easily delineated for exclusion from plantation activity. Photographs taken just before clear-felling give a clear view of the existing vegetation and suitability of site for a particular species can be judged from it. The photographs can also be used to assess percentage success of plantations. Thus, it is extremely useful in organizing plantation activity and assessing its result.

ii) Forest protection and damage control

Forests are exposed to various threats to ***man, insects, fungi, and some adverse climatic factors.*** Aerial photographs are extremely useful not only for assessing the damage caused by these sources but also giving timely information for effective prevention and counter measures. ***Colour infra-red photographs are particularly good for this type of job.***

In recent years ***encroachments*** have increased mani-folds. The aerial photographs taken at intervals can show the location and extent of encroachments so that prosecution the offenders can take place expeditiously. Comparison of such photographs taken at interval of five or ten years can show the rate of loss of forest area from this as well as other causes.

Apart from this no reliable statistics about area under shifting cultivation and forest areas which are under constant threat to forest fires. Aerial photographs provide one of the easiest and quickest methods of supplying this information accurately so that remedial measures can be planned.

iii) Wildlife conservation and management

Aerial photography has been now using for census of certain animals and habitat/cover mapping of national parks. e.g.,

1. Census of brow antlered deer and hog deer using aerial photography

   In Keibul Lanjao Sanctuary, Manipur, 1975 and 1979

2. Census of Wild Ass using aerial photography

   In Runn of Kutch, Gujarat

3. Habitat/cover map using aerial photography

   In Kanha National Park, Madhya Pradesh

iv) Soil conservation and watershed management

Aerial photography is very useful in soil conservation, as it helps in early detection of its causes, and planning remedial measures. With the development of river valley projects, watershed management has assumed great importance and aerial photography can give precise information of the catchment as soon as any river valley project is planned so that by the time the project can be completed, the catchment is properly afforested to prevent silting up of the dam.

Another aspect in which aerial photography is used, when planning a river valley project is the assessment of land use pattern in the area of submergence, and in case of forest area, planning its systematic exploitation while in case of agricultural land, planning for the rehabilitation of displaced persons else where.

v) Recreation resource inventory

There is an increasing demand for development of new recreation areas. Aerial photography is exceedingly useful for this work because it can facilitate: i) selection of new recreation areas from amongst several alternative proposals, and ii) Intensive management of existing sites by supplying information on terrain analysis and by providing cover type maps and similar other information.

vi) Development of communication

Aerial photography is very useful in preliminary reconnaissance of the area for alignment of road and later in selection of final alignment. Use of aerial photography for the purpose saves both time and money.

In short, ***aerial photography, besides being a powerful tool for preparation of forest inventory, in assessing of timber and other forest resources, is extremely useful in managing forests on multiple-use basis*** and more recently in creating an environmental awareness in the general public.

Other roles of aerial photography in forestry

Aerial photography is used :

1. To measure the tree and stand characteristics
   - ❐ Tree and stand height
   - ❐ Crown density
   - ❐ Crown count
   - ❐ Crown diameter
   - ❐ Tree and stand volume
2. Area measurement
3. For interpretation of land use classes
4. Soil and erosion surveys
5. Interpretation of forest cover types
6. Preparation of different types of maps
7. Forest inventory

- ❐ Stratification
- ❐ Sampling designs in forest inventories
- ❐ Assessment of wood resources in non-forest lands

Apart from this Aerial photography is also used in :

- ❐ Stock mapping for working plan or working scheme
- ❐ Survey and demarcation of the protected un classed forests

- Forest protection and damage control
- Assessment of damage by shifting cultivation and encroachments
- Assessment of area and productivity of forests
- Assessment of production of various minor forest products per unit area
- Wildlife, nature conservation and recreation
- Site classification
- Soil conservation, management of forest watersheds

Remote sensing and GIS in Karnataka

Realizing the capabilities of eye in the sky- satellite remote sensing in addressing various issues of natural resources planning, development and management in an unbiased manner Govt. of Karnataka established **Karnataka State Remote Sensing Applications Centre (KSRSAC)** in the year 1989. Over the last one and half decade, KSRSAC has been responsible for undertaking wide range of projects, such as :

1. Watershed planning and management
2. Land use/land cover planning and management
3. Crop acreage estimation
4. Waste land mapping
5. Vegetation cover mapping and change detection
6. Environmental impact assessment
7. Ground water prospect mapping
8. Command area mapping
9. Coastal zone change studies
10. Urban and regional planning

Objectives of KSRSAC

- Inventory and monitoring of natural resources
- Under take short and long-term remote sensing and GIS programmes in association with line departments
- Regulate and monitor RS and GIS activities in the state as nodal agency.

- Acquire and maintain a library of spatial database of Karnataka
- Provide training for the officers of the line departments
- Organize workshops, seminars, symposia and awareness programmes
- Dissemination of information
- Interface with academy and industry

Milestones

1. KSRSAC was conceptualized in the year ***1986***
2. Operationalised in the year 1989 as the major Govt. Deptt. under Deptt. of Science and Technology, Govt. of Karnataka
3. Transferred to Planning Deptt. in 1992
4. Made as an autonomous organization in 1997 under Planning Deptt.
5. Transferred to Deptt. of Information Technology in 1999 and is being continued as an autonomous centre.

Administrative set up

President	:	Chief Secretary of the government
Executive committee	:	Chaired by Secretary to Government
Member secretary	:	KSRSAC Director
Members	:	Principal secretaries/secretaries of the line departments

❑❑❑

Chapter 10

Basics of Geographic Information System (GIS) & Global Positioning System (GPS)

A Geographic Information System (GIS) integrates hardware, software, and data for capturing, managing, analyzing, and displaying all forms of geographically referenced information. GIS also allows the integration of these data sets for deriving meaningful information and outputting the information derivatives in map format or tabular format.

Three Views of a GIS

A GIS can be viewed in three ways:

1. **The Database View** : A GIS is a unique kind of database of the world—a geographic database (geo database). It is an "Information System for Geography." Fundamentally, a GIS is based on a structured database that describes the world in geographic terms.
2. **The Map View :** A GIS is a set of intelligent maps and other views that show features and feature relationships on the earth's surface. Maps of the underlying geographic information can be constructed and used as "windows into the database" to support queries, analysis, and editing of the information.
3. **The Model View :** A GIS is a set of information transformation tools that derive new geographic datasets from existing datasets. These geo-processing functions take information from existing datasets, apply analytic functions, and write results into new derived datasets.

By combining data and applying some analytic rules, we can create a model that helps to answer the question you have posed.

Global Positioning System (GPS)

The Global Positioning System (GPS) is a space-based global navigation satellite system (GNSS) that provides reliable location and time information in all weather and at all times and anywhere on or near the Earth, when and where there is an unobstructed line of sight to four or more GPS satellites. It is maintained by the United States government and is freely accessible by anyone with a GPS receiver. GPS was created and realized by the U.S. Department of Defence (USDOD) and was originally run with 24 satellites. It was established in 1973 to overcome the limitations of previous navigation systems.

Basic concept of GPS

A GPS receiver calculates its position by precisely timing the signals sent by GPS satellites high above the Earth. Each satellite continually transmits messages that include :

a) The time the message was transmitted

b) Precise orbital information (the ephemeris)

c) The general system health and rough orbits of all GPS satellites (the almanac).

The receiver uses the messages it receives to determine the transit time of each message and computes the distance to each satellite. These distances along with the satellites' locations are used with the possible aid of trilateration, depending on which algorithm is used, to compute the position of the receiver. This position is then displayed, perhaps with a moving map display or latitude and longitude; elevation information may be included. Many GPS units show derived information, such as direction and speed, calculated from position changes. Three satellites might seem enough to solve for position since space has three dimensions and a position near the Earth's surface can be assumed. However, even a very small clock error multiplied by the very large speed of light, the speed at which satellite signals propagate results in a large positional error. Therefore, receivers use four or more satellites to solve for the receiver's location and time. The very accurately computed time is effectively hidden by most GPS applications, which use only the location. A few specialized GPS applications do, however, use the time; these include time transfer, traffic signal timing, and synchronization of cell phone base stations. Although four satellites are required for normal operation, fewer apply in special cases. If one variable is already known, a receiver can determine its position using only three satellites. For example, a ship or aircraft may have known elevation. Some GPS receivers may use additional clues or assumptions (such as reusing the last known altitude, dead reckoning, inertial navigation, or including information from the vehicle computer) to give a less accurate (degraded) position, when fewer than four satellites are visible.

1. Application of Remote Sensing and GIS in Civil Engineering

Remote sensing and GIS techniques become potential and indispensable tools for solving many problems of civil engineering. Remote sensing observations provides data on the earth's resources in a spatial format, GIS co-relates different kinds of spatial data and their attribute data, so as to use them in various fields of civil engineering.

a) In structural engineering

Structural Health Monitoring (SHM) provides designers with feedback of structural performance, assisting in development of structures with higher utility and lower manufacturing costs. Structural Health Monitoring nowadays continues to advance from conventional strain gauges to FBG Fibre Optic Sensors (FOS) and major breakthroughs in wireless remote monitoring. Fibre optic sensors use optical wavelength of fibre Bragg grating to measure temperature and strain. FOS has many advantages over the traditional electrical system, such as:

- Suitable for long-term permanent SHM: monitor structure during construction stage and whole lifespan as well
- No calibration needed
- One cable can have hundreds of the sensors
- Simple installation
- Cable can run kilometres, no length limit
- Fibre optic sensors use light signal - no electrical sparking, intrinsically safe
- Gauge length can be few metres long to measure global behaviours of structures
- Suitable for both static and dynamic measurement

The primary of monitoring is to ensure the longevity and safety of the structure as well as optimizing its management. To implement corrective measures and maintenance action, monitoring must be enable the timely detection of any condition or behaviour that could deteriorate the structure, deem it unsafe or potentially results in its failure.

The monitoring programme plays a fundamental role during the construction phase, as it enables the verification of design hypotheses and construction processes, affecting, in some cases, the construction rate of the structures and overall quality. Most defects are introduced already at the time of construction. Monitoring also allows performance evaluation of new materials and technologies used in bridge construction and rehabilitation. This objective is easily achieved with fibre optic sensors since these sensors effectively integrate in new materials, such as fibre reinforced polymer composite.

Furthermore, fibre optic sensors adapt perfectly to long-term monitoring of bridges behaviour as well as short-term monitoring of bridges dynamic behaviour under traffic load. Finally, monitoring can be used as a tool for supervised lifetime extension of bridges approaching the end of their life or in need of major repair. It ensures that such bridges are operated safely while allowing the postponement of major investments and traffic disruption.

b) Town Planning and Urban Development

To achieve the objectives of making metropolis cities more livable and of international standard, a coordinated and integrated approach among the various agencies involved in urban development and provision of services are needed including participatory process in planning and implementation at local body levels. As well as to have planned and organized disposal of population through growth centres, which will acts as counter-magnets to the cities growth. This growth may not able to withstand the existing infrastructure, traffic, road, drainage and utility networks, etc. Advance urban planning is required for a planned development of the area for which up-to-date real time and accurate information are the vital important. Geographical Information system & Remote Sensing is inevitable technology in the development of national Infrastructure and planning and they provide solution related to many environmental.

Applications of Remote Sensing to Hydrology and Hydrogeology

The Hydrological Cycle

A brief overview of hydrological processes will help to set a framework for describing those areas where remote sensing can assist in observing and in managing water resource system. Generally speaking, the hydrological cycle traces water through different physical processes, from liquid water through evaporation into the atmosphere, back into the liquid (or sometimes the frozen) state as precipitation falling on land areas either run off into rivers and streams, or percolate into the soil, or evaporate. Moisture reaching the water table becomes ground water. As a general rule, both surface and ground water flow under the force of gravity towards streams and lakes, and ultimately oceans. The return of water to the oceans can thought of as completing the cycle.

Precipitation

Accurate measurement of precipitation is a continuing goal in meteorological research and a continuing need in hydrology which depends greatly on these data for modelling. Ground-based radar is probably the most accurate method of determining a real precipitation in use today. Satellite images from GOES, NOAA, TIROS-N, TRMM and NIMBUS opened a whole new world of data on clouds and frontal systems. Work carried out by several researchers has led to the following conclusions:

a) In thick clouds (more than one kilometre) rain is possible when the upper surface of the cloud is at less that –15°C.

b) The probability of rain is inversely proportional to the temperature of the upper surface of the cloud.

c) Precipitation intensity is directly proportional to the area of the upper surface of the cloud at temperature of less than –15°C.

Snow

For the hydrologists who must forecast water levels, snow represents one of the most complicated and most difficult to measure parameters. Snow extent, distribution, water equivalent, water content, thickness and density, all play a large part in assessment of the snow-pack' s contribution to runoff. Snow pack water equivalent has been measured by aircraft gamma-radiation surveys in the USA. The method is based on the absorption of natural gamma radiation by water (snow). As hydrologists come to accept satellite remote-sensing data on snow mapping, they also come to learn the limitations of satellite remote sensing. Despite some indications that the reflectance of snow may, under certain circumstances, be related to the snow thickness.

Glaciers : Glaciers play an important role in the hydrological cycle of many mountainous areas. Terrestrial photography of glaciers was an important early reference method. Traversing and conducting scientific studies on glaciers are difficult, and glacieologists were quick to appreciate the value of remote sensing, first from aircraft, later from satellites (Landsat, HCMM, NIMBUS and IceSat ..., etc).

Surface Water : One of the best known applications of remote sensing to water resources is the inventorying of surface water bodies, particularly streams, lakes, marshes and bogs, within a given region. The area covered by open water is readily delineated by various remote-sensing techniques because of the particular radiation characteristics of water. Decreased reflectivity of soils moisturized at the surface facilities the delineation of recently flooded areas, if these are barren. The delineation of floods in vegetation-covered areas is more difficult, but is possible either by use of radar or through a combination of radiation and topographic data. Remotely sensed data obtained on flood-plain characteristics can be combined with data obtained during floods for flood mapping and delineating flood hazard areas. Characteristics of river channel, such as width, depth, roughness, degree of tortuousity and braiding can also be obtained from remote-sensing surveys.

Ground Water

Ground water is concerned with water in the saturated zones beneath the surface of the Earth. Ground water information most useful to water resource managers includes: the presence or absence of ground water in designated areas, the depth to ground water, the quantity and quality of water available for development, recharge rates to aquifer, the possible impact of pumping on land subsidence, a real extent of the aquifer, locations of recharge and discharge areas, and the interaction between withdrawals at wells and natural discharge into rivers. Whereas, this information is generally sought by hydrogeologists using conventional methods, remote sensing can help in the planning of conventional measurements and can be used to estimate some hydrogeological variables quantitatively and others qualitatively. The storage capacity of ground water reservoirs depends on their extent, which depends on geological properties of the area. Ground water forms the base flow for many streams and is the source of water for springs and seeps.

Applications in Hydrology

Hydrology is the study of water on the Earth's surface, whether flowing above ground, frozen in ice or snow, or retained by soil. Hydrology is inherently related to many other applications of remote sensing, particularly forestry, agriculture and land cover, since water is a vital component in each of these disciplines. Most hydrological processes are dynamic, not only between years, but also within and between seasons, and therefore, require frequent observations. Remote sensing offers a synoptic view of the spatial distribution and dynamics of hydrological phenomena, often unattainable by traditional ground surveys. Radar has brought a new dimension to hydrological studies with its active sensing capabilities, allowing the time window of image acquisition to include inclement weather conditions or seasonal or diurnal darkness.

Examples of hydrological applications include :

- Wetlands mapping and monitoring,
- Soil moisture estimation,
- Snow pack monitoring/delineation of extent,
- Measuring snow thickness,
- Determining snow-water equivalent,
- River and lake ice monitoring,
- Flood mapping and monitoring,
- Glacier dynamics monitoring (surges, ablation)
- River/delta change detection
- Drainage basin mapping and watershed modelling
- Irrigation canal leakage detection
- Irrigation scheduling

Applications of Remote Sensing in Weather Forecasting and Warnings

A. Applications of meteorological satellites

Meteorological satellites are indispensable in weather forecasting and warning services. Because of their huge areal coverage, meteorological satellite images can be used to keep track of weather systems days before they come close to an area. This is particularly useful in monitoring severe weather systems like tropical cyclones. The very basic application of meteorological satellite is in identification of clouds. Clouds can be broadly classified into three categories according to the cloud base height, namely low, medium and high clouds. Some clouds, such as cumulonimbus (a type of thundery clouds), span the three layers. Different clouds have different characteristics in terms of shape and pattern and have different tones in the visible and infrared images. These differences enable the identification of clouds using a combination of the visible and the infrared images. For instance, fog and low dense clouds are characterized by their sharp boundary and smooth texture on satellite image. They appear in bright white to medium gray tone on the visible image, but in dark to medium grey colour on infrared image. Thundery clouds, such as cumulonimbus, however, contains abundant moisture and extends to great height. They appear in globular shape and are in very bright tone on both the visible and infrared images. Apart from identification of clouds, meteorological satellites are widely used in many areas of applications. Here below are some examples:

An excellent tool in unravelling volcanic ash beneath clouds. The operating principle is that volcanic ash and clouds exhibit different characteristics in the IR1 and IR2 infrared images.

Remote sensing application in geomorphology

Geomorphology is the science of study of the landforms of the earth Geomorphological analysis of surface forms of the earth is a direct form of interpretation from space images. Aerial photos with required forward overlap usually provide the third dimension of height, which adds to the precision of interpretation including morphometry. Geomorphology as a science developed much later than geology although several aspects of geomorphology are embedded in geological processes.

Geomorphology deals with the genesis of relief forms of the surface of the earth's crust. Certain natural processes are responsible for the forms of the surface of the earth. A thorough understanding of various processes leading to landforms is necessary to understand the environment in which we live. Remote sensing is an effective tool in this understanding, as aerospace images contain integrated information of all that is on the ground, the landform, the ecology, the resources contained in the area and the impact of human actions on the natural landscape. The dynamism with which changes occur in the landscape is brought out effectively by repeated coverage of images of the same area at different times. Images convey many things even to the untrained eye and for a professional, it conveys much more including many features hitherto unknown or unseen on the ground.

Geomorphology - basic concepts

The earth's surface forms are primarily due to hypogene or endogenous processes, which include diastrophism, leading to geologic structure, tectonic activity and volcanism leading to volcanic landforms. These forms are modified by epigene or exogenous processes, which include erosion and depositional activities of water, wind and ice. Other activities include weathering, mass wasting or movement of material by gravitational action, land-ocean interaction resulting in landforms due to waves, currents, tides and tsunamis. Climate is another important factor, which has relevance in shaping of the earth's surface because the processes that act upon the surface material are different in different climatic zones (Van Westen, 1994).

For example, limestone forms hills in a dry climate whereas in wet climate, it forms Karst topography with sink holes, caves and caverns pre-domination.

Remote Sensing applications in Agriculture

Introduction

Agriculture resources are among the most important renewable, dynamic natural resources. Comprehensive, reliable and timely information on agricultural resources is very much necessary for a country like India whose mainstay of the economy is agriculture. Agriculture survey are presently conducted throughout the nation in order to gather information and associated statistics on crops, rangeland, livestock and other related agricultural resources. These information of data are most importance for the implementation of effective management decisions at local, panchayat and district levels. In fact, agricultural survey is a backbone of planning and allocation of the limited resources to different sectors of the economy. With increasing population pressure throughout the nation and the concomitant need for increased agricultural production (food and fibre crops as well as livestock), there is a definite need for improved management of the nation agricultural resources. In order to accomplish this, it is first necessary to obtain reliable data on not only the types, but also the quality, quantity and location of these resources.

Remote sensing and its Importance in Agricultural survey

Remote sensing is nothing but a means to get the reliable information about an object without being in physical contact with the object. It is on the observation of an object by a device separated from it by some distance utilizing the characteristics response of different objects to emissions in the electromagnetic energy is measured in a number of spectral bands for the purpose of identification of the object. In such study single tabular form of data or map data is not sufficient enough which can provide can be, combined with information's obtained from existing maps and tabular data :

a) Remote Sensing techniques using various plate form has provide its utility in agricultural survey.

b) Satellite data provides the actual synoptic view of large are at a time, which is not possible from conventional survey methods.

c) The process of data acquisition and analysis is very fast through Geographic Information System (GIS), as compared to conventional methods.

Remote Sensing techniques have a unique capability of recording data in visible as well as invisible (i.e. ultraviolet, reflected infrared, thermal infrared and microwave, etc.) part of electromagnetic spectrum. Therefore, certain phenomenon, which cannot be seen by human eye, can be observed through remote sensing techniques i.e. the trees, which are affected by disease, or insect attack can be detected by remote sensing techniques much before human eyes see them.

Present system of Generating agricultural data and its Problems

The present system of agricultural data is collected throughout the nation. The main responsibility of collection agricultural survey lies on the Director of Land Records, Director of Agriculture and District Statistical Office under the Ministry of Agriculture. These data are collected not only on a local but also some extent of district and state level. The associate of agricultural survey on crops (crop production, type of crop and crop yield), range land (condition of range, forest type, water quality, types of irrigation system and soil characteristics) and livestock (livestock population, sex of animal, types of farm and distribution of animals).

The basic problems in this survey are :

1. Reliability of data
2. Cost and benefits
3. Timeless
4. Incomplete sample frame and sample size
5. Methods of selection
6. Measurement of area
7. Non-sampling errors
8. Gap in geographical coverage
9. Non-availability of statistics at disaggregated level.

Remote Sensing techniques make it use before the remote sensing data may provide solution to these particular problems of agricultural survey.

Advantages of Remote Sensing techniques in Agricultural Survey

With the primary aim of improving the present means of generating agricultural data, a number of specific advantages may result form the use of remote sensing techniques :

1. **Vantage point :** Because the agricultural landscape depends upon the sun as a source of energy, it is exposed to the aerial view and, consequently, is ideally suited or remote sensing techniques.

2. **Coverage :** With the use of high-altitude sensor platforms, it is now possible to record extensive areas on a single image. The advent of high-flying aircraft and satellites, single high quality images covering thousand of square miles.
3. **Permanent record :** After an image is obtained, it serves as a permanent record of a landscape at a point in time which agriculture changes can be monitored and evaluated.
4. **Mapping Base :** Certain types of remote sensing imagery are, in essence, pictorial maps of the landscape and after rectification (if needed), allow for precise measurement (such as field acreages) to be made on the imagery, obviating time-consuming on the ground surveys. These images may also aid ground data sampling by serving as a base map for location agriculture features while in the field, and also as a base for the selection of ground sampling point or areas.
5. **Cost savings :** The costs are relatively small, when compared with the benefits, which can be obtained form interpretation of satellite imagery.
6. **Real-time capability :** The rapidly with which imagery can be obtained and interpreted may help to eliminate the lock of timeliness which plagues, so many agricultural survey.

Other advantages of Remote Sensing

1. Easy data acquisition over inaccessible area.
2. Data acquisition at different scales and resolutions
3. The images are analyzed in the laboratory, thus reducing the amount of fieldwork.
4. Colour composites can be produced from three individual band images, which provide better details of the area then a single band image or aerial photograph.
5. Stereo-satellite data may be used for three-dimensional studies. At present, all advantages listed above have been demonstrated either operationally or experimentally:

Application of Remote sensing techniques for Agricultural survey

The specific application of remote sensing techniques can be used for:

i) Detection
ii) Identification
iii) Measurement
iv) Monitoring of agricultural phenomena

Area of specific applications

a) Applicable to crop survey

1. Crop identification	2. Crop acreage
3. Crop vigor	4. Crop density
5. Crop maturity	6. Growth rates
7. Yield forecasting	8. Actual yield
9. Soil fertility	10. Effects of fertilizer
11. Soil toxicity	12. Soil moisture
13. Water quality	14. Irrigation requirement
15. Insect infestations	16. Disease infestations
17. Water availability	18. Location of canals

b) Applicable to range survey

1. Delineation of forest types	2. Condition of range
3. Carrying capacity	4. Forage
5. Time of seasonal change	6. Water quality
7. Soil fertility	8. Soil moisture
9. Insect infestations	10. Wildlife inventory
12. Location of water	

c) Applicable to livestock survey

1. Cattle population	2. Sheep population
3. Pig population	4. Poultry Population
5. Age sex distribution	6. Distribution of animals
7. Animal behaviour	8. Disease identification
9. Types of farm buildings	

Application of remote sensing in Seismology

A wide range of satellite methods is applied now in seismology. The first applications of satellite data for earthquake exploration were initiated in the 70s, when active faults were mapped on satellite images. It was a pure and simple extrapolation of airphoto geological interpretation methods into space. The modern embodiment of this method is alignment analysis. Time series of alignments on the Earth's surface are investigated before and after the earthquake. A further application of satellite data in seismology is related with geophysical methods. Electromagnetic methods have about the same long history of application for seismology. Stable statistical estimations of ionosphere-lithosphere relation were obtained based on satellite ionozonds. The most successful current project "DEMETER" shows impressive results. Satellite thermal infra-red data were applied for earthquake research in the next step. Numerous results have confirmed previous observations of thermal anomalies on the Earth's surface prior to earthquakes. A modern trend is the application of the outgoing long-wave radiation for earthquake research. Spectacular pictures of co-seismic deformations were presented. Current researches are moving in the direction of pre-earthquake deformation detection. GPS technology is also widely used in seismology both for ionosphere sounding and for ground movement detection. Satellite gravimetry has demonstrated its first very impressive results on the example of the catastrophic Indonesian earthquake in 2004. Relatively new applications of remote sensing for seismology as atmospheric sounding, gas observations, and clouds analysis are considered as possible candidates for applications.

Introduction

Remote sensing has been used for earthquake research from the 70s, with the first appearance of satellite images. First of all, it was used in structural geological and geomorphological research. Active faults and structures were mapped on the base of satellite images. This method is very limited in time series analysis. There was no possibility to measure short-term processes before and after the earthquake. It was simple an extrapolation of airphoto geological interpretation methods into space. The modern version of this method is active tectonic analysis with the application of alignment analysis. Time series of alignment distributions on the Earth's surface are investigated before and after an earthquake.

The current situation of remote sensing application for earthquake research indicates a few phenomena, related with earthquakes, particularly the Earth's surface deformation, surface temperature and humidity, atmosphere temperature and humidity, gas and aerosol content. Both horizontal and vertical deformations scaled from tens of centimetres to metres are recorded after the shock. Such deformations are recorded by the Interferometric Synthetic Aperture Radar (InSAR) technique with confidence. Pre-earthquake deformations are rather small, on the order of centimetres. A few cases of deformation mapping before the shock using satellite data are known at present time. Future developments lay in precision longwave SAR systems with medium spatial resolution and combined with the GPS technique. There are numerous observations of surface and near surface temperature increases of 3–5 °Ñ prior to the Earth crust earthquakes. Methods of earthquake prediction are developing using thermal infrared (TIR) surveys. Multiple evidence of gas and aerosol content changes before earthquakes are reported for ground observations. Satellite methods allow one to measure the concentrations of gases in atmosphere: O_3, CH_4, CO_2, CO, H_2S, SO_2, HCl and aerosols. However, the spatial resolution and sensitivity of modern systems restricts the application of satellite gas observation in seismology and the first promising results have been obtained only for ozone, aerosol and air humidity.

Deformations

One of the main directions of remote sensing application for seismology is deformation mapping. Surface deformations in seismic cycles can be divided into three phases: pre-seismic or inter-seismic, co-seismic and post-seismic ones. Co-seismic deformations are evaluated upto metres and tens of metres while pre-seismic movements amount to centimetres. Post-seismic deformations are also measured in centimetres, but subsequent landslides can increase deformations to meters. Most current research is focused on co-seismic and post-seismic (landslide) deformations.

Discussion

A wide spectrum of satellite remote sensing methods is applied in seismology now-a-days. The value of these methods for earthquake research is varied. Optical methods have limited applications, mostly for rapid assessment of damages in an epicentral zone. Other applications, such as alignment analysis and clouds form analysis related with earthquakes do not have an adequate scientific basis for seismological application. Vigorous extension of InSAR methods applications in seismology is observed now. Modern radar systems in conjuncttion with GPS/GLONASS will provide whole seismic cycle monitoring.

Broad application of In SAR methods is limited by the high data cost and complex data analysis. Thermal satellite data applications are developing in two directions at the moment: thermal anomalies in seismic fault research and emitted longwave radiation measurements in seismic zones. Thermal anomalies research in seismic faults is developing in the direction of seismic activity monitoring and close integration with ground observations. Emitted long wave radiation observations demonstrate promising results, but data accumulation is required. The nature of ongoing longwave radiation anomalies remains unclear.

Some common remarks on satellite data application in seismology can be made:

1. The level of automatic data processing is insufficient. There is still too much manual labour and author arbitrariness in data processing—this concerns both exotic earthquake cloud analysis and high technique radar methods. Some results are irreproducible.
2. There is a weak physical and geological basis for many of the proposed methods. The nature and driving forces of some phenomena need clarification and connection with current understanding of physics and geology.

❑❑❑

Chapter 11

Applications in Forestry

(Status monitoring, forest fire damage assessments, soil survey and mapping, vegetation cover mapping, species identification, height and volume estimation, wildlife habitat suitability mapping)

Background : On March 24, 2011, multiple wildfires began across south-east Georgia. Strong, gusty winds and low relative humidity created extreme fire behavior and dangerous firefighting conditions. By March 28, 2011, a state of emergency had been issued by Governor Nathan Deal for five counties in South Georgia (Coffee, Bacon, Ware, Clinch, and Long Counties), and over 43,500 acres was lost due to wildfire. The four largest fires in counties under a state of emergency were: Sessom Community Fire (Bacon/Ware Counties; 20,597 acres), Arabia Bay Fire (Clinch County; 11,673 acres), Elim Church Road Fire (Long County; 4,029 acres), and Mosley Road Fire (Coffee County; 1,418 acres). Long-term drought and extreme fire weather conditions resulted in limited control and an estimated timber mortality rate of 70–90%.

The intensity and extent of damage resulting from each of these fires was more severe than damage sustained during the massive Waycross wildfires of 2007.

An initial assessment was made by the Georgia Forestry Commission on Sessom Community Fire, Arabia Bay Fire, Elim Church Road Fire, and Mosley Road Fire to determine fire severity and possibly predict survival of the damaged stands. Based upon previous research done in the 1960's, 1980's and 1990's on wildfires in Florida, stands were rated for damage and classified in one of three categories: **Light, Moderate, and Severe**. Although one might assume that pine bark beetles automatically play a decimating role following fires of this magnitude, this has not always been the case, as documented in the 1990's research (Hanula, Meeker, Barnard, *et al.*). The 1980's research conducted by Dixon (*et al.*) did indicate an increase in Ips Engraver

Wild Life Habitat Suitability Mapping

Wildfire Damage Assessment

for the 2011 Southeast Complex Fires

Chip Bates & Mark McClure, Forest Health Management

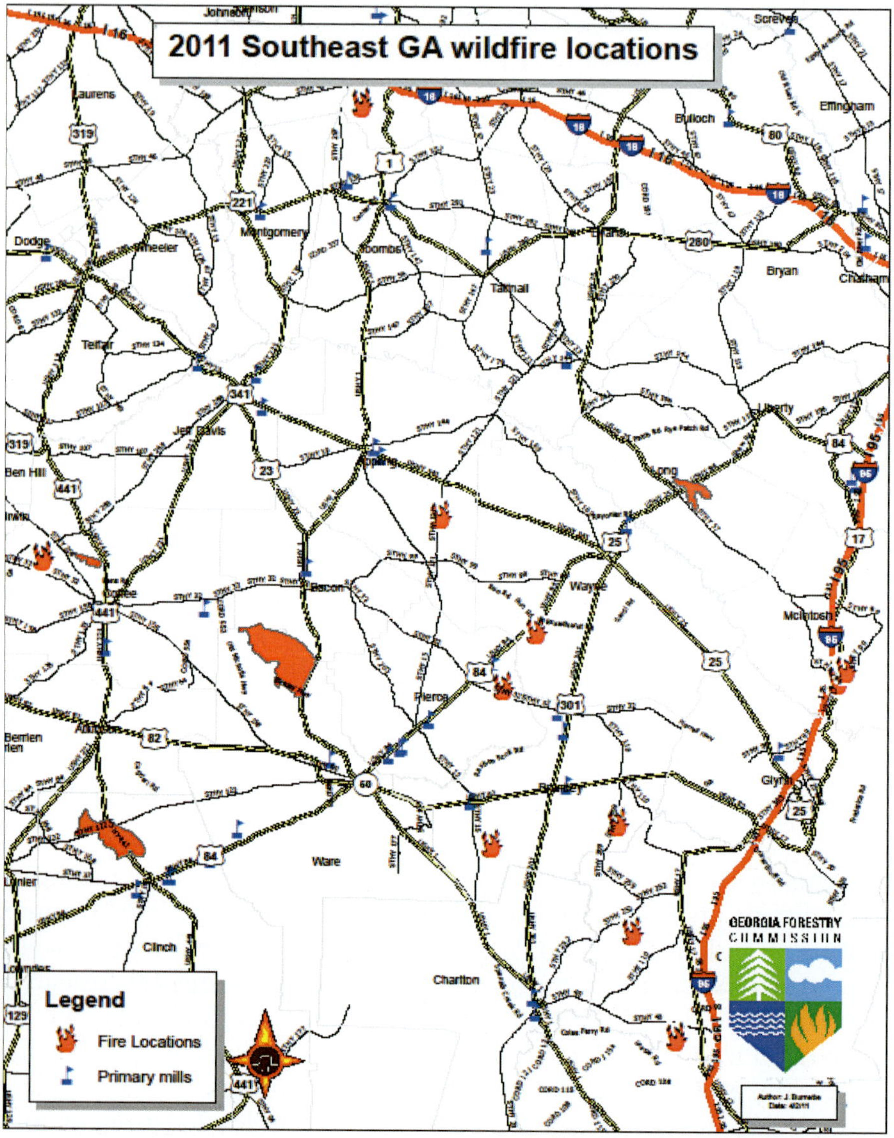

Beetles for upto two growing seasons, following an April wildfire. One fact that is certain is that catastrophic wildfires often coincide with long term drought, and both of these factors cause tree stress, decline of growth and mortality. There are numerous factors to consider when assessing fire impact, and pine bark beetles are just one. Other stand factors that were observed are: *Root Damage, Basal Stem Damage, Stem Char and Severity, Crown Consumption and Scorch, Appearance and Damage of Growing Shoots (if present), and overall appearance of the stands.*

Survey Methods : Wildfires burn with different levels of intensities. They are impacted by time of day, daily weather patterns, and fuel composition. These variations result in a mosaic of burn patterns across the landscape, variations within each stand and from section to section. Survey efforts sought to classify each stand for its overall appearance and damage levels. It should be noted that this is not an exact science, but rather an overall indicator or forest health. In marginal areas, future weather patterns, particularly rainfall, will play a major role in tree survival throughout the remainder of the 2011 growing season.

Fires burned through most sampled areas two weeks prior to this survey. Evaluations were made primarily along travel corridors in an effort to get reliable samples and to make worthwhile observations and predictions. A great deal of timber salvage has begun, but it is only a small percentage of what needs to occur.

Timber Evaluation Protocol : Most observers evaluate the severity of a burn by looking at the canopy of the stand or tree. A thorough and more accurate evaluation should be conducted by using a systematic approach which forecasts potential insect and disease problems and associated tree mortality. This process begins with the root system and litter layer, followed by examination of the bole for stem damage and char, and then upto the crown, inspecting for crown consumption and scorch, damage to shoot growth, and the overall appearance of the stand. Any sign or symptom of insect presence must be noted. A post-fire evaluation guidelines form is available to document this process.

Root Damage : Most tree roots can be found in the top six inches of soil, with many roots positioned at the soil surface. As fire burns the litter and duff layer, tree roots are exposed and feeder roots are killed. When fire burns into the ground, organic material is removed, which produces a depression around the tree's base - **"The Halo of Death"**

"Halo of Death"

Stem Damage and Char : Stem char refers to the blackened portion of the stem that was exposed to direct flame. After needles drop and crown scorch can't be effectively measured, the stem char line remains as a permanent record indicating fire severity on the tree.

Stem damage is a good visual representation of the intensity and duration of the fire around the stem of the tree. Complete charring on all sides of the stem or noticeable bark consumption are indicators of possible injury to the cambium layer. Normally, clear sap flowing between the bark plates is evidence of cambial damage, and it is highly attractive to insects. The higher the bark char the greater the potential for mortality.

"Bark Char"

"Stem Damage and Char"

Crown Consumption and Scorch: Crown scorch refers to needles that weren't directly burned in the fire, but were exposed to lethal temperatures during the fire event. These needles will typically drop from the tree within 30 days of turning reddish - brown. Crown consumption indicates the portion of the crown/living needles that were directly burned off by the fire. These stands should be harvested immediately.

"Crown Consumption"

Terminal Bud Damage and Mortality: Heat, of varying levels, over 140 degrees for more than one minute is lethal to most plant tissues. In the early spring to summer, candling terminal buds are highly susceptible to damage from excessive heat and are a key identification point in determining tree survival.

"Shepherd's Crook"

Crown scorch does not automatically indicate mortality of the tree. Surrounding needles can protect the terminal bud from severe scorch or consumption providing a chance for survival. Normally a significant portion of green canopy will be noted at the top of the tree and the terminal bud will be straight and unburned. If the terminal bud is "Shepherd's Crooked", there is little to no chance for survival.

Insect Sign and Symptoms: Southern Bark beetles and Ambrosia beetles are attracted to stressed and fire-damaged stands. Any indication of boring dust or insect entry holes should be noted.

Hardwood Damage : Hardwoods also face serious damage from fire. Many Southern hardwood species have thin bark and shallow root systems, making them more susceptible to fire damage and eventual death from the direct exposure to flames and heat. Fire will not normally burn in the lower, wet areas where we find hardwood stands, but the extended drought conditions have produced lower moisture levels and dry organic material that can be several inches to multiple feet deep. Hardwood stands with heavy fuel build-up around the bases of trees experienced long burn periods that may have damaged the stem bases and root system. These damaged trees may experience more complications from root and fungal disease and insects.

Survey Findings

Using the 2007 Waycross Complex Fires as a reference; the Southeast Complex fires in Coffee, Bacon, Ware, Clinch, and Long Counties were more severe in damage and are expected to have a greater mortality rate. The "Big Turnaround Fire" in 2007 had an overall mortality rate of 60-70 % with a severity rating of 52%.

The 2011 Southeast Complex fires have an overall expected mortality rate approximately 20% - 25% higher than the 2007 Waycross Complex.

Damage Estimate

Damage Estimate	Arabia Bay	Elim Church Rd.	Mosely Rd.	Sessom Comm.
Severe	84%	63%	83%	68%
Moderate	13%	17%	14%	10%
Light	3%	20%	3%	22%

- Stands rated as **severe** had extreme damage, and a significant portion of the trees are already dead or will not likely survive this growing season.
- **Moderate** stands had significant damage, but a portion of the trees likely will survive this growing season. The resulting stand, however, may not be fully stocked and some landowners should liquidate these areas.
- **Light** stands had relatively minor damage, and a majority of trees within these areas will survive. Some mortality has or will likely occur within these areas also, but it is expected to be relatively minor and no management action should be needed.

Overall : A mortality rate of **90%** could be applied to the Arabia Bay and Mosely Road fires, while a mortality rate of **70-80%** could be applied to the Elim Church Road and Sessom Community fires.

Product Class	Arabia Bay	Elim Church Rd.	Mosely Rd.	Sessom Comm.
Pre-merchantable	29%	9%	2%	23%
Pulpwood	44%	39%	15%	48%
Chip-n-Saw	10%	22%	58%	19%
Sawtimber	11%	20%	20%	6%
Hardwood Drain	6%	10%	5%	4%

Almost all areas other than Hardwood Drains were pine plantations, so the above product classes refer to one of the pine species listed below :

Species	Arabia Bay	Elim Church Rd.	Mosely Rd.	Sessom Comm.
Loblolly Pine	35%	32%	28%	35%
Slash Pine	47%	57%	66%	60%
Longleaf Pine	0%	1%	1%	1%
Hardwood/Cypress	18%	10%	5%	4%

Variation in Fire Damage within Timber Stands and Across the Landscape

Severe fire activity and crown consumption can be seen in the black areas

Brown needles indicate crown scorch and less severe damage

Green canopy indicates the lowest levels of damage (if little to no basal stem or root damage occurred).

Two weeks post-burn. Stem char is approximately 75%, crown scorch is close to 100%. This stand would be rated as "severely" damaged. Note that when all needles fall, the potential for re-burn exists.

Two weeks post-burn. Crown was completely consumed by fire. Obviously, this stand is "severely" damaged and should be harvested immediately.

Younger stands tended to suffer more severe damage, if they were greater than 3 years old (4 years old stand at left)

Stands that had heavy fuel buildup around the bases of trees (right) experienced long burn periods that may have damaged the stem bases and roots. This has been indicated as a mortality factor in the past.

Salvage harvest operations began immediately after the fire

Remote Sensing Systems

Lidar

LIDAR or **LADAR** {**Light(Laser) Detection And Ranging**} is an optical remote sensing technology that can measure the distance to, or other properties of a target by illuminating the target with light, often using pulses from a laser. LIDAR technology has application in archaeology, geography, geology, geomorphology, seismology, forestry, remote sensing and atmospheric physics.

General description

LIDAR uses ultraviolet, visible, or near infrared light to image objects and can be used with a wide range of targets, including non-metallic objects, rocks, rain, chemical compounds, aerosols, clouds and even single molecules. A narrow laser beam can be used to map physical features with very high resolution.

LIDAR has been used extensively for atmospheric research and meteorology. Downward-looking LIDAR instruments fitted to aircraft and satellites are used for surveying and mapping.

Wavelengths in a range from about 10 micrometres to the UV (250 nm) are used to suit the target. Typically light is reflected via backscattering. Different types of scattering are used for different LIDAR applications, most common are Rayleigh scattering, Mie scattering and Raman scattering as well as fluorescence.

Suitable combinations of wavelengths can allow for remote mapping of atmospheric contents by looking for wavelength-dependent changes in the intensity of the returned signal.

Basic of LIDAR System

LIDAR system involves a laser range finder reflected by a rotating mirror (top). The laser is scanned around the scene being digitized, in one or two dimensions gathering distance measurements at specified angle intervals.

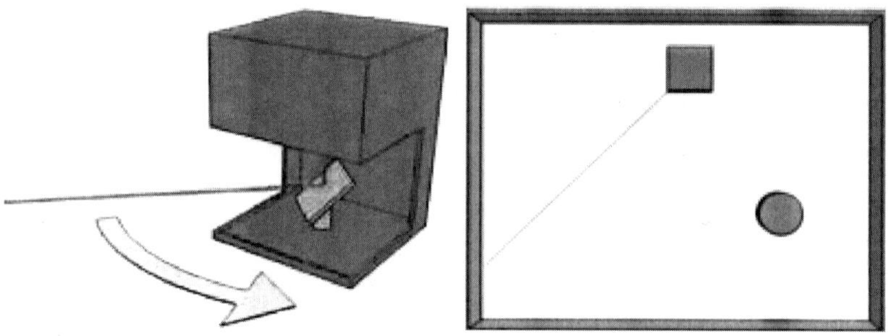

In general, there are two kinds of LIDAR detection schema :

i) **Incoherent :** As direct energy detection (which is principally an amplitude measurement) and

ii) **Coherent detection:** (which is best for doppler, or phase sensitive measurements). Coherent systems generally use Optical detection which being more sensitive than direct detection allows them to operate a much lower power but at the expense of more complex transceiver requirements.

In both coherent and incoherent LIDAR, there are two types of pulse models:

a) ***Micropulse*** **systems** and

b) ***High energy*** **systems.**

a) **Micropulse systems** have developed as a result of the ever increasing amount of computer power available combined with advances in laser technology. They use considerably less energy in the laser, typically on the order of one microjoule, and are often "eye-safe," meaning they can be used without safety precautions.

b) **High-power systems** are common in atmospheric research, where they are widely used for measuring many atmospheric parameters: the height, layering and densities of clouds, cloud particle properties (extinction coefficient, backscatter coefficient, depolarization), temperature, pressure, wind, humidity, trace gas concentration (ozone, methane, nitrous oxide, etc.).

Major components of LIDAR system

1. **Laser** : 600–1000 nm lasers are most common for non-scientific applications. They are inexpensive but since they can be focused and easily absorbed by the eye the maximum power is limited by the need to make them eye-safe. A common alternative 1550 nm lasers are eye-safe at much higher power levels since this wavelength is not focused by the eye, but the detector technology is less advanced and so, these wavelengths are generally used at longer ranges and lower accuracies. They are also used for military applications as **1550 nm is not visible** in night vision goggles unlike the shorter 1000 nm infrared laser.

Airborne topographic mapping LIDAR generally use 1064 nm diode pumped YAG lasers, while bathymetric systems generally use 532 nm frequency doubled diode pumped YAG lasers because 532 nm penetrates water with much less attenuation than does 1064 nm. Laser settings include the laser repetition rate (which controls the data collection speed). Pulse length is generally an attribute of the laser cavity length, the number of passes required through the gain material (YAG, YLF, etc.), and Q-switch speed. Better target resolution is achieved with shorter pulses, provided the LIDAR receiver detectors and electronics have sufficient bandwidth.

2. **Scanner and optics** : How fast images can be developed is also affected by the speed at which it can be scanned into the system. There are several options to scan the azimuth and elevation, including dual oscillating plane mirrors, a combination with a polygon mirror, a dual axis scanner. Optic choices affect the

angular resolution and range that can be detected. A hole mirror or a beam splitter are options to collect a return signal.

3. **Photodetector and receiver electronics** : Two main photodetector technologies are used in LIDAR: solid state photodetectors, such as silicon avalanche photodiodes, or photomultipliers. The sensitivity of the receiver is another parameter that has to be balanced in a LIDAR design.
4. **Position and navigation systems** : LIDAR sensors that are mounted on mobile platforms, such as airplanes or satellites require instrumentation to determine the absolute position and orientation of the sensor. Such devices generally include a Global Positioning System receiver and an Inertial Measurement Unit (IMU).

Applications of LIDAR

This LIDAR-equipped mobile robot uses its LIDAR to construct a map and avoid obstacles.

1. Agriculture

Agricultural Research Service scientists have developed a way to incorporate LIDAR with yield rates on agricultural fields. This technology will help farmers direct their resources towards the high-yield sections of their land. LIDAR also can be used to help farmers determine which areas of their fields to apply costly fertilizer. LIDAR can create a topological map of the fields and reveals the slopes and sun exposure of the farm land. Researchers at the Agricultural Research Service blended this topological information with the farm land's yield results from previous years. This technology is valuable to farmers because it indicates which areas to apply the expensive fertilizers to achieve the highest crop yield.

2. Archaeology

LIDAR has many applications in the field of archaeology including aiding in the planning of field campaigns, mapping features beneath forest canopy, and providing an overview of broad, continuous features that may be indistinguishable on the ground. LIDAR can also provide archaeologists with the ability to create high-resolution digital elevation models of archaeological sites that can reveal micro-topography that are otherwise hidden by vegetation. LIDAR-derived products can be easily integrated into a Geographic Information System (GIS) for analysis and interpretation. With LIDAR, the ability to produce high-resolution datasets quickly and relatively cheaply can be an advantage. Beyond efficiency, its ability to penetrate forest canopy has led to the discovery of features that were not distinguishable through traditional geospatial methods and are difficult to reach through field surveys.

3. Biology and conservation

LIDAR has also found many applications in forestry. Canopy heights, biomass measurements, and leaf area can all be studied using airborne LIDAR systems. Similarly, LIDAR is also used by many industries, including Energy and Railroad, and the Department of Transportation as a faster way of surveying. Topographic maps can also be generated readily from LIDAR, including for recreational use, such as in the production of orienteering maps. In oceanography, LIDAR is used for estimation of phytoplankton fluorescence and generally biomass in the surface layers of the ocean. Another application is airborne lidar bathymetry of sea areas too shallow for hydrographic vessels.

4. Geology and soil science

High-resolution digital elevation maps generated by airborne and stationary LIDAR have led to significant advances in geomorphology, the branch of geoscience concerned with the origin and evolution of Earth's surface topography. LIDAR's abilities to detect subtle topographic features, such as river terraces and river channel banks, measure the land surface elevation beneath the vegetation canopy, better resolve spatial derivatives of elevation, and detect elevation changes between repeat surveys have enabled many novel studies of the physical and chemical processes that shape landscapes. Aircraft-based LIDAR and GPS have evolved into an important tool for detecting faults and measuring uplift. The output of the two technologies can produce extremely accurate elevation models for terrain that can even measure ground elevation through trees.

5. Hydrology

LIDAR offers a lot of information to the aquatic sciences. High-resolution digital elevation maps generated by airborne and stationary LIDAR have led to significant advances in the field of Hydrology.

6. Meteorology and atmospheric environment

The first LIDAR systems were used for studies of atmospheric composition, structure, clouds, and aerosols. Initially based on ruby lasers, LIDAR for meteorological applications was constructed shortly after the invention of the laser and represent one of the first applications of laser technology. *Elastic backscatter LIDAR* is the simplest type of lidar and is typically used for studies of aerosols and clouds. The backscattered wavelength is identical to the transmitted wavelength, and the magnitude of the received signal at a given range depends on the backscatter coefficient of scatterers at that range and the extinction coefficients of the scatterers along the path to that range. The extinction coefficient is typically the quantity of interest. *Differential Absorption LIDAR* (DIAL) is used for range-resolved measurements of a particular gas in the atmosphere, such as ozone, carbon dioxide, or water vapour. The LIDAR transmits two wavelengths: an "on-line" wavelength that is absorbed by the gas of interest and an off-line wavelength that is not absorbed. The differential absorption between the two wavelengths is a measure of the concentration of the gas as a function of range. DIAL LIDARs are essentially dual-wavelength backscatter LIDARS. *Raman LIDAR* is also used for measuring the concentration of atmospheric gases, but can also be used to retrieve aerosol parameters as well. Raman LIDAR exploits inelastic scattering to single out the gas of interest from all other atmospheric constituents. A small portion of the energy of the transmitted light is deposited in the gas during the scattering process, which shifts the scattered light to a longer wavelength by an amount that is unique to the species of interest. *Doppler LIDAR* is used to measure wind speed along the beam by measuring the frequency shift of the backscattered light. Scanning LIDARs, have been used to measure atmospheric wind velocity in a large three-dimensional cone.

7. Law enforcement

LIDAR speed guns are used by the police to measure the speed of vehicles for speed limit enforcement purposes and offer a number of advantages over radar speed guns.

8. Military

Few military applications are known to be in place and are classified, but a considerable amount of research is underway in their use for imaging. Higher resolution systems collect enough detail to identify targets, such as tanks. Examples of military applications of LIDAR include the Airborne Laser Mine Detection System (ALMDS) for counter-mine warfare by Arete Associates. Utilizing LIDAR and THz interferometry wide area raman spectroscopy, it is possible to detect chemical, nuclear, or biological threats at a great distance. In atmospheric physics, LIDAR is used as a remote detection instrument to measure densities of certain constituents of the middle and upper atmosphere, such as potassium, sodium, or molecular nitrogen and oxygen. These measurements can be used to calculate temperatures. LIDAR can also be used to measure wind speed and to provide information about vertical distribution of the aerosol particles. In nuclear fusion research facility, LIDAR Thomson Scattering is used to determine Electron Density and Temperature profiles of the plasma.

9. Robotics

LIDAR technology is being used in Robotics for the perception of the environment as well as object classification. Refer to the Military section above for further examples.

10. Transportation

LIDAR has been used in Adaptive Cruise Control (ACC) systems for automobiles. Systems use a lidar device mounted on the front of the vehicle, such as the bumper, to monitor the distance between the vehicle and any vehicle in front of it. In the event the vehicle in front slows down or is too close, the ACC applies the brakes to slow the vehicle. When the road ahead is clear, the ACC allows the vehicle to accelerate to a speed preset by the driver. Refer to the Military section above for further examples.

❑❑❑

Chapter 12

Fundamentals of Geoinformatics and its Applications

Introduction

The art, science and technology dealing with the acquisition, storage, processing, production, presentation and dissemination of geoinformation is called geoinformatics. According to Burrough (1987), it is a powerful set of tools for collecting, storing, retrieving at will, transforming and displaying of spatial data form the real world. Geoinformatics is made up of Geographical Information System (GIS), Remote Sensing (RS) and Global Positioning System (GPS). It is the advancing science and technology, which can be used for research and development in any discipline. For a geographer, it is a powerful tool to generate and geocoded data and attribute database to help taking decisions appropriately for planning. Geographical Information System (GIS) is a computer system capable of capturing, storing, analyzing, and displaying geographically referenced information i.e. data identified according to location. Parker (1988) defined GIS as "an information technology which stores, analyses, and display both spatial and non-spatial data.

Objectives of GIS

- Optimization of efficiency of planning and decision-making.
- Capacity to integrate information from various sources.
- Provide efficient means for data distribution and handing.
- Elimination of redundant data base.
- Complex analysis/query involving geographical referenced data to generate new information.

Elements of GIS

- **Hardware** : It is the physical unit (electronic machine) that ultimately performs the job. Monitor, printer, plotter, scanner digitizer, etc.
- **Software** : A set of programs written to perform a specific job with the help of hardware.

Operating system: windows 2000, 2007, Windows XP, Windows Vista, Unix, etc.

GIS Software: Arcinfo, AraGIS, Arcview, Erdas, GRAM, GRAM++, Idrisi, ILWIS, Autocadmap, etc.

- **Data** : Two types of data :

  a) Spatial data and

  b) Non-spatial (attribute) data.
- **Liveware** : Users or people.
- Data Model : A set of logical definitions for characterizing the geographical data in order to present spatial information and their attributes, the two approaches for data model are:

1. Raster representation and
2. Vector representation

Followings are few raster file formats commonly used in geoinformatics:

- **ADRG -** National Geospatial-Intelligence Agency (NGA)'s ARC Digitized Raster Graphics
- **BIL** - Band Interleaved by Line (image format linked with satellite derived imagery)
- **CADRG** - National Geospatial-Intelligence Agency (NGA)'s Compressed ARC Digitised Raster Graphics (nominal compression of 55:1 over ADRG)
- **ECRG** - National Geospatial-Intelligence Agency (NGA)'s Enhanced Compressed ARC Raster Graphics (Better resolution than CADRG and no colour loss)
- CIB - National Geospatial-Intelligence Agency (NGA)'s Controlled Image Base (type of Raster Product Format)

- Digital raster graphic (DRG) - digital scan of a paper USGS topographic map
- ECW - Enhanced Compressed Wavelet (from ERMapper). A compressed wavelet format, often lossy.
- ESRI grid - proprietary binary and metadataless ASCII raster formats used by ESRI
- GeoTIFF - TIFF variant enriched with GIS relevant metadata
- IMG - ERDAS IMAGINE image file format
- JPEG2000 - Open-source raster format. A compressed format, allows both lossy and lossless compression.
- MrSID - Multi-Resolution Seamless Image Database (by Lizardtech). A compressed wavelet format, often lossy.
- netCDF-CF - netCDF file format with CF medata conventions for earth science data. Binary storage in open format with optional compression. Allows for direct web-access of subsets/ aggregations of maps through OPenDAP protocol. And Followings are few raster file formats commonly used in geoinformatics.
- Geography Markup Language (GML) – XML based open standard (by OpenGIS) for GIS data exchange
- AutoCAD DXF - Contour elevation plots in AutoCAD DXF format
- Shapefile (SHP)- ESRI's open, hybrid vector data format using SHP, SHX and DBF files
- Simple Features - Open Geospatial Consortium specification for vector data
- MapInfo TAB format - MapInfo's vector data format using TAB, DAT, ID and MAP files
- National Transfer Format (NTF) – National Transfer Format (mostly used by the UK Ordnance Survey)
- TIGER - Topologically Integrated Geographic Encoding and Referencing
- Cartesian coordinate system (XYZ) - Simple point cloud
- Vector Product Format – National Geospatial-Intelligence Agency (NGA)'s format of vectored data for large geographic databases.

- GeoMedia - Intergraph's Microsoft Access based format for spatial vector storage.
- ISFC - Intergraph's Micro Station based CAD solution attaching vector elements to a relational Microsoft Access database
- Personal Geodatabase - ESRI's closed, integrated vector data storage strategy using Microsoft's Access MDB format
- File Geodatabase - ESRI's file-based geodatabase format, stored as folders in a file system. ESRI also has an enterprise Geodatabase format for use in an RDBMS.
- Coverage - ESRI's closed, hybrid vector data storage strategy. Legacy ArcGIS Workstation/ArcInfo format with reduced support in ArcGIS Desktop lineup
- Spatial Data File - Autodesk's high performance geodatabase format, native to MapGuide
- GeoJSON - A light weight format based on JSON, used by many open source GIS packages
- SOSI Standard - A spatial data format used for all public exchange of spatial data in Norway Grid formats (for elevation)
- USGS DEM - The USGS' Digital Elevation Model
- DTED - National Geospatial-Intelligence Agency (NGA)'s Digital Terrain Elevation Data
- GTOPO30 - Large complete Earth elevation model at 30 arc seconds
- SDTS - The USGS' successor to DEM Other formats
- Binary Terrain - The Virtual Terrain Project's Binary Terrain format
- Dual Independent Map Encoding (DIME) – A historic GIS file format, developed in the 1960s
- Well-known text (WKT) – ASCII spatial projection description (ESRI uses a *.prj extension)
- Well-known binary (WKB) - Binary spatial projection description
- World file - Geo-referencing a raster image file (e.g. JPEG, BMP)

Need of GIS

Collector, SP, Administrators, Academicians- University, MLA & MP, Environmentalist, Agriculturist, Disaster management cell e.g. For agricultural management GIS has to be considered multi-dimensional with attribute dimension, spatial dimension &temporal dimension GIS offers capabilities of integrating multisector, multilevel and multiperiod database.

GIS satisfy the following specific needs

1. The ability to preprocess data from large stores into a form suitable for analysis, including operation, such as reformatting, change of projection, re-sampling and generalizations.
2. Direct support for analysis and modelling, so that from of analysis, calibration of models, forecasting, and prediction are all handled through instruction to the GIS.
3. Post processing of result including such operations as reformatting, tabulation, report generation, and mapping.

An information system is the chain of operations-planning, observation and collection of data- storage and analysis of data- Decision-making process.

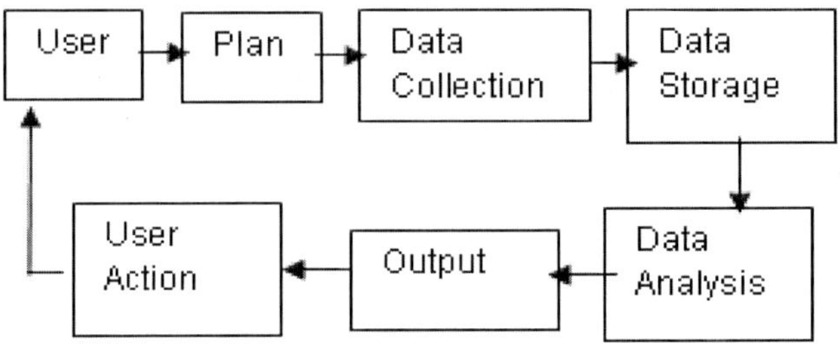

Fig. 1 Decision-making process

Four key activities can be enhanced by using GIS

- Measurement
- Monitoring
- Mapping
- Modeling

Contributing disciplines to GIS.

- Geography
- Cartography
- Computer software
- Surveying
- Photogrammetry
- Remote Sensing Technology
- Mathematics/Statistics

Steps in GIS analysis

- Importing Data
- Georeferencing
- Digitizing and editing
- Data analysis:

  a) Network analysis

  b) Digital Elevation Model (DEM).

  c) Morphometric analysis, etc.
- Query Building.

Remote Sensing (RS) means acquiring information about a phenomena or object or surface while at distance from it (without being physical contact with it).

Principles of Remote Sensing

Different objects return different amount and kind of energy in different bands of the electromagnetic spectrum incident upon it. This property of the object depends on structural, physical and chemical composition, surface roughness, angle of incidence, intensity and wavelength of radiant energy, hence we can identify various objects by collecting and analyzing returned energy.

Stages in RS

- Emission of energy or EMR (Sun)
- Transmission of energy from the source to the surface of the earth as well as adsorption and scattering of energy in the atmosphere.
- Interaction of EMR with the earth surface Reflection.
- Transmission of the energy from the surface to the remote sensor.
- Sensor data output.
- Data transmission, processing and analysis.

Platforms and Sensors

Platform is a stage to mount the camera or sensor to acquire the information about a target. Based on their altitude above the earth surface they are Ground borne. Air borne and Space borne. Sensor is a device that gathers energy, converts it into a signal and presents it in a suitable form for obtaining information about the target under investigation.

Two types of sensors : 1. Active and 2. Passive.

Resolution

Resolution is defined as the ability of the system to render the information at the smallest discretely separable quantity in terms of distance (spatial), wavelength band of EMR (spectral), time (temporal) and radiation quantity (radiometric).

Image Interpretation

An interpreter can derive lot of information by studying the quantitative and qualitative aspects of images and photographs. An image taken from air or space is a pictorial presentation of the pattern of a landscape.

Two types of image interpretation

1. Visual Image Interpretation: Elements are Shape, Size, Tone, Shadow, Pattern, Texture, Location, Resolution, and Stereoscopic Appearance.

2. Digital Image Processing (DIP): It can be categorized into the following types of computer—assisted operations:
   a) Image Rectification: aim at correcting distorted data
   b) Image Enhancement: increase the amount of information that can be visually interpreted form the data. e.g. Contrast, Edge, Band Ratioing.
   c) Image Classification: Automatically categorize all pixels in an image into land cover classes or themes.

Two Types : (i) Unsupervised and (ii) Supervised classification: Ground truth sites should be large enough, well-distributed and readily identifiable on the toposheets and the image.

Global Positioning System (GPS) also called as NAVSTAR GPS (Navigational System with Time and Ranging) was created by the U.S. Department of Defense for defense reasons. This system is being used for many civil applications now-a-days. It is a system of earth-orbiting satellites transmitting precisely timed signals (a similar system of Russian Federation: GLONASS - global navigation satellite system) which provides direct measurement of position on the Earth's surface and the location is expressed in latitude/longitude. The GPS receiver measures the travel time of signals transmitted from triangulated satellites. It calculates the satellites' latitude, longitude, altitude and speed. GPS provides an address to every square metre of the earth's surface.

The Five Main Points of Global Positioning System

1. The basis of the GPS system is triangulation from the satellites. Triangulation is when distance is measured using the travel time of radio messages from satellite.
2. Two receivers are needed; the first receiver acts as a "reference point". It is stationary and monitors, transmits and records the corrections to errors. The second receiver is placed in a position that is recorded and calculates position of satellite (with data). Then it compares the answer with a known position; the difference is the error in the GPS signal.
3. The satellite and the receiver generate the same pseudo-random code at the same time. The distance to a satellite is determined by measuring the time it takes for a radio signal to reach us from that satellite.

4. This code allows for the GPS system to work with very low power signals and small antenna.
5. Distance to satellites is measured by accurate timing, using atomic clocks on board. Four satellite range measurements will reduce errors in timing.

Three major segments

i) **Space Segment**: Consists of 24 satellites at an altitude of 20200 km above the earth surface.

ii) **Ground control segment:** Five ground stations distributed around the earth for monitoring and controlling the satellite system continuously

iii) **The user segment:** It includes all those who use GPS tracking equipment to receive GPS signals to satisfy positioning requirements.

Applications of Geoinformatics

Geoinformatics has become a popular tool in the recent studies in geography, environmental science, town planning, defense studies, etc. Some active domains for application of Geoinformatics are:

The main producers and sources

Topographical Mapping

National Mapping Agencies, private Mapping Companies Land Registration and Cadastre.

Hydro graphic Mapping

Military Organizations. Remote Sensing companies and satellite agencies. Natural resource surveys; Geologists; Hydrologists; Physical Geographers and Soil Scientists; Land Evaluators; Ecologists and Biogeographers; Meteorologists and Climatologists; Oceanographers.

The main types of geographical data available

Topographic maps at a wide range of scales Satellite and airborne scanner images and Photographs Administrative boundaries: Census tracts and census data; Postcode areas Statistical data on people, land cover, land use at a wide range of levels.

Data from marketing surveys

Data on utilities (gas, water, electricity lines, cables) and their locations. Data on rocks, water, soil, atmosphere, biological activity, natural hazards, and disasters collected for a wide range of spatial and temporal levels of resolution.

Some current applications

1. Agriculture Monitoring and management from farm to national levels.
2. Archaeology site description and scenario evaluation.
3. Environment Monitoring, modelling, and management for land degradation; land evaluation and rural planning; landslides; desertification; water quality and quantity; plagues; air quality; weather and climate modelling and prediction.
4. Epidemiology and health location of disease in relation to environmental factors.
5. Forestry Management, planning, and optimizing extraction and replanting.
6. Emergency services Optimizing fire, police, and ambulance routing; improved understanding of crime and its location.
7. Navigation Air, sea, and land.
8. Marketing Site location and target groups; optimizing goods delivery.
9. Real Estate Legal aspects of the cadastre, property values in relation to location, insurance.
10. Regional/local planning Development of plans, costing, maintenance, management. Road and railway planning and management.
11. Site evaluation and costing. Cut and fill, computing volumes of materials.
12. Social studies Analysis of demographic movements and developments.
13. Tourism Location and management of facilities and attractions. Utilities Location, management, and planning of water, drains, gas, electricity, telephone, cable services.

The various phenomenon on the earth can be studied using geoinformatics, as river morphology.

River Morphology

River morphology, primarily, deals with the formation, of channels and their characteristics. Morphological studies, therefore, include plan form, length and width of the channels, slope and sediments distribution along with the rivers, cross sectional geometry, sediment loads in relation to discharge and river stability. The investigations are more aimed at to find out causes, which are responsible for behaviour of a river in a particular manner. The studies also deal with the causes and sites of aggradations and degradation along the river course. Erosion caused by the rivers is one of major problems of rivers flowing through alluvial plains. River morphological observation of help in identification of areas undergoing serve bank erosion and taking appropriate river engineering measures to combat bank erosion. Construction of spurs, embankment, dykes, etc. are primarily dependent on the river behaviour- its dynamic nature. River morphological studies are of prime requisite while planning for construction of dam, barrage, culverts and so on. Selection of sites for sediments gauge stations needs morphological information of the rivers.

Information needs

The data is available, in general, in the form of maps, reports, charts, tables, etc. Use of maps in river basins studies is indispensable. The maps quite often are available in different formats and scales. Most crucial component of the river basin studies is a reliable, comprehensive, compatible and recent data on various watershed characteristics. Most of the data available today on the network.

Conclusion

Geoinformatics has become an inevitable tool in recent times for research, planning and projections of various aspects of different fields of studies.

References

1. Begni G. Escadafal R. Fontannaz D. and Hong-Nga Nguyen A.-T. (2005). Remote sensing: a tool to monitor and assess desertification. Les dossiers thématiques du CSFD. Issue 2. 44 pp.

2. C. L. Parkinson, A. Ward, M. D. King (Eds.) Earth Science Reference Handbook - A Guide to NASA's Earth Science Program and Earth Observing Satellite Missions, National Aeronautics and Space Administration Washington, D. C.

3. Chang, K. T. (1989). "A comparison of techniques for calculating gradient and aspect from a gridded digital elevation model". International Journal of Geographical Information Science **3** (4): 323–334.

4. *Chang, K. T. (2008). Introduction to Geographical Information Systems. New York: McGraw Hill. p. 184.*

5. Clarke, K. C., 1986. Advances in geographic information systems, computers, environment and urban systems, Vol. 10, pp. 175–184.

6. Ditter, R., Haspel, M., Jahn, M., Kollar, I., Siegmund, A., Viehrig, K., Volz, D., Siegmund, A. (2012) Geospatial technologies in school – theoretical concept and practical implementation in K-12 schools. In: International Journal of Data Mining, Modelling and Management (IJDMMM): Future GIS: Riding the Wave of a Growing Geospatial Technology Literate Society; Vol. X

7. *Fitzgerald, Joseph H.* "Map Printing Methods". *Archived from* the original *on 2007-06-04.* Retrieved 2007-06-09.

8. Fu, P., and J. Sun. 2010. Web GIS: Principles and Applications. ESRI Press. Redlands, CA. ISBN 1-58948-245-X.

9. *Goodchild, Michael F (2010). "Twenty years of progress: GIScience in 2010". Journal of Spatial Information Science.*

10. GRAS-SAF (2009), Product User Manual, GRAS Satellite Application Facility, Version 1.2.1, 31 March, 2009.

11. *Greene, R.; Devillers, R.; Luther, J.E.; Eddy, B.G. (2011). "GIS-based multi-criteria analysis". Geography Compass* ***5/6****: 412–432.*

12. *Grigoriev À.N. (2015).* "Method of radiometric distortion correction of multi-spectral data for the earth remote sensing.". *Scientific and Technical Journal of Information Technologies, Mechanics and Optics 15 (4): 595–602.*

13. *Guo, Huadong; Huang, Qingni; Li, Xinwu; Sun, Zhongchang; Zhang, Ying (2013).* "Spatiotemporal analysis of urban environment based on the vegetation-impervious surface-soil model" *(Full text article available).* Journal of Applied Remote Sensing *8: 084597.*

14. Heywood, I., Cornelius, S., Carver, S. (2006). An Introduction to Geographical Information Systems (3rd ed.).

15. *Howard, A., et al., (Aug 19, 2015). "Remote sensing and habitat mapping for bearded capuchin monkeys (Sapajus libidinosus): landscapes for the use of stone tools". Journal of Applied Remote Sensing 9 (1).*

16. *Hunter G. J.; Goodchild M. F. (1997).* "Modelling the uncertainty of slope and aspect estimates derived from spatial databases" *(PDF). Geographical Analysis* **29** *(1): 35–49.*

17. Jones, K.H. (1998). "A comparison of algorithms used to compute hill slope as a property of the DEM". Computers and Geosciences **24** (4): 315–323.

18. K. Calvert, J. M. Pearce, W.E. Mabee, "Toward renewable energy geo-information infrastructures: Applications of GI Science and remote sensing that can build institutional capacity" Renewable and Sustainable Energy Reviews **18**, pp. 416–429 (2013). open access

19. *Liu, Jian Guo & Mason, Philippa J. (2009).* Essential Image Processing for GIS and Remote Sensing. *Wiley-Blackwell. p. 4.*

20. *Longley, P. A.; Goodchild, M. F.; McGuire, D. J.; Rhind, D. W. (2005). Analysis of errors of derived slope and aspect related to DEM data properties. Geographic Information Systems and Science (West Sussex, England: John Wiley and Sons). p. 328.*

21. *Lovison-Golob, Lucia.* "Howard T. Fisher". *Harvard University.* Retrieved 2007-06-09.

22. *Ma, Y.; Guo, Y.; Tian, X.; Ghanem, M. (2011). "Distributed Clustering-Based Aggregation Algorithm for Spatial Correlated Sensor Networks". IEEE Sensors Journal* **11** *(3): 641.*

23. *Maliene V, Grigonis V, Palevièius V, Griffiths S (2011).* "Geographic information system: Old principles with new capabilities". *Urban Design International. pp. 1–6.*

24. Mills, J.P.; et al. (1997). "Photogrammetry from Archived Digital Imagery for Seal Monitoring". The Photogrammetric Record 15 (89): 715–724.

25. NASA (1986), Report of the EOS data panel, Earth Observing System, Data and Information System, Data Panel Report, Vol. IIa., NASA Technical Memorandum 87777, June 1986, 62 pp.

26. Schott, John Robert (2007). Remote sensing: the image chain approach (2nd ed.). Oxford University Press. p. 1.

27. Schowengerdt, Robert A. (2007). Remote sensing: models and methods for image processing (3rd ed.). Academic Press. p. 2.

28. Stewart, J.E.; et al. (2014). *"Finescale ecological niche modelling provides evidence that lactating gray seals (Halichoerus grypus) prefer access to fresh water in order to drink"*. Marine Mammal Science 30 (4): 1456–1472.

29. Stork, E.J., Sakamoto, S.O., and Cowan, R.M. (1999) "The integration of science explorations through the use of earth images in middle school curriculum", Proc. IEEE Trans. Geosci. Remote Sensing 37, 1801–1817

30. Tim Foresman 1997 The History of GIS (Geographic Information Systems): Perspectives from the Pioneers. (Prentice Hall Series in Geographic Information Science) Prentice Hall PTR; 1st edition (November 10, 1997), 416 p.

31. Twiss, S.D.; et al. (2001). *"Topographic spatial characterisation of grey seal Halichoerus grypus breeding habitat at a sub-seal size spatial grain"*. Ecography 24 (3): 257–266.

32. Winther, Rasmus G. (2014) "Mapping Kinds in GIS and Cartography" in Natural Kinds and Classification in Scientific Practice, edited by C. Kendig

33. Zhou, Q.; Liu, X. (2003). *"Analysis of errors of derived slope and aspect related to DEM data properties"*. Computers and Geosciences 30: 269–378.

❑❑❑

Chapter 13

Model Questions Bank

Q I. Choose the Correct Answer for the Following

1. Deforestation is occurring in many countries and it is, therefore, difficult to detect the ongoing damage. Which of the following types of remote sensing would be best suited for locating deforestation?

   a) Thermal infrared　　b) Microwave

   c) Radar　　d) Sonar

   e) Colour infrared

2. Multi-spectral remote sensing allows researchers to obtain data about the Earth in areas of the electromagnetic spectrum beyond what the eye can see. If you were studying the effects of drought on vegetation, which portion of the electromagnetic spectrum would provide the most pertinent data?

   a) Thermal infrared　　b) Middle infrared

   c) Near infrared　　d) Blue spectrum

   e) Red spectrum

3. You need to find the scale of a map have been given for a class project. In order to do so, you decide to measure a distance between two real world features and relate that to the distance shown on the map. You find that the distance between two buildings is 400 feet. On the map that same distance is 3 inches. What is the scale of the map?

   a) 1:1,200　　b) 1:16,000

   c) 1:1,600　　d) 1:100

   e) 1:133.33

5. Using the latitude and longitude system (degrees, minutes, seconds), 20 minutes is equal to what?

   a) 1/3 hour
   b) 0.20°
   c) 200 seconds
   d) 1/20°
   e) 1/3°

6. An oblate ellipsoid is what?

   a) A circle rotated about its major axis
   b) A circle rotated about an ellipse
   c) An ellipse rotated about a circle
   d) An ellipse rotated about its longer axis
   e) An ellipse rotated about its shorter axis

7. To compare, overlay, or cross-analyze two maps in a GIS ____

   a) Both maps must be in digital form
   b) Both maps must be in the same map projection
   c) Both maps must be at the same equivalent scale
   d) Both maps must be on the same coordinate system
   e) All of the above

8. Which of the following types of remote sensing would be most useful in obtaining an accurate terrain representation of Venus?

   a) Radar
   b) Microwave
   c) Landsat
   d) Sonar
   e) Thermal infrared

9. An automated system for the capture, storage, retrieval, analysis, and display of spatial data is known as ________

   a) GPS
   b) Landsat
   c) GIS
   d) Stereoscope

10. Living vegetation appears ________ on false-colour IR images.

    a) White
    b) Black
    c) Blue
    d) Red

11. Precise measurement of Earth features can be obtained from ___
    a) High-oblique photographs
    b) Low-oblique photographs
    c) Vertical aerial photograph
    d) All the above types of aerial photo
    e) Only high- and low-oblique aerial photographs
12. Lines connecting points of equal air temperature are known as ___________
    a) Isohyets
    b) Isotherms
    c) Isobars
    d) Contour lines
13. You measure the length of a particular stream section on a topographic map with a scale of 1:24000 is 3.2 inches. The real-world length of the stream is ___________
    a) 1.45 miles
    b) 14.55 miles
    c) 1.21 miles
    d) 12.1 miles
14. The largest scale of the following is ___________
    a) 1:24000
    b) 1:62500
    c) 1:100000
    d) 1:500000
    e) 1:1000000
15. With which type of remote sensing imagery would a baseball field of artificial turf be differentiated from natural grass?
    a) Radar imagery
    b) Colour infrared
    c) Black and white photography
    d) Colour photography
16. A line connecting points of equal precipitation is
    a) An isobar
    b) A contour line
    c) An isohyet
    d) An isotherm
17. When topographic maps produced by the United State Geological Survey have been updated, the revisions are done in ___________
    a) Black
    b) White
    c) Green
    d) Blue
    e) Purple

18. Which of the following remote sensing technologies uses sound?

    a) Radar b) Colour infrared imaging

    c) Thermal infrared imaging d) Sonar

    e) Microwave sensing

19. They may also be used to detect the ________ of various chemicals, providing data on chemical concentrations in the atmosphere.

    a) Spectroscopy

    b) Mössbauer spectroscopy

    c) Atomic emission spectroscopy

    d) Emission spectrum

20. Stereographic pairs of ________ have often been used to make topographic maps by imagery and terrain analysts in trafficability and highway departments for potential routes.

    a) Photography b) Aerial photography

    c) Digital camera d) Cloudscape photography

21. Radiometers and ________ are the most common instrument in use, collecting reflected and emitted radiation in a wide range of frequencies.

    a) Infrared b) Photometer

    c) Absorption spectroscopy d) Spectroscopy

22. ________ is an example of active remote sensing where the time delay between emission and return is measured, establishing the location, height, speed and direction of an object.

    a) Ultra high frequency b) X band

    c) Radar d) Weather radar

23. Conventional ________ is mostly associated with aerial traffic control, early warning, and certain large scale meteorological data.

    a) Radar b) Weather radar

    c) X band d) Ultra high frequency

24. Passive; _______ is used for detecting, ranging and measurements of underwater objects and terrain.

 a) Echo sounding

 b) Sonar

 c) Marine mammals and sonar

 d) Underwater acoustics

25. This data revealed minute perturbations in the Earth's gravitational field _______ that may be used to determine changes in the mass distribution of the Earth, which in turn may be used for geological or hydrological studies.

 a) World Geodetic System b) Geology

 c) Longitude d) Geodesy

26. Seismograms taken at different locations can locate and measure _______ (after they occur) by comparing the relative intensity and precise timing.

 a) Earthquake b) Geotechnical engineering

 c) S-wave d) Landslide

27. Sensors include cameras, digital scanners and _______, while platforms usually consist of aircraft and satellites.

 a) LIDAR b) Lightning detection

 c) Meteorology d) Weather radar

28. Technically, a GIS is a system that includes mapping software and its application to _______, remote sensing, land surveying, mathematics, photogrammetry, geography, and tools that can be implemented with GIS software.

 a) Computer graphics

 b) Spatial analysis

 c) Geographic information system

 d) Cartography

29. _______ is the design and production of maps, or visual representations of spatial data.

 a) Cartography b) Spatial analysis

 c) Geography d) Geographic information system

30. ________ data can be directly entered into a GIS from digital data collection systems on survey instruments using a technique called Coordinate Geometry (COGO).

    a) Inventions in medieval Islam

    b) Dam

    c) Surveying

    d) Civil engineering

31. ________ is the process by which a surface is created, usually a raster dataset, through the input of data collected at a number of sample points.

    a) Polynomial
    b) Numerical analysis
    c) Linear interpolation
    d) Interpolation

32. The GIS display depends on the viewing point of the ________ and time of day of the display, to properly render the shadows created by the sun's rays at that latitude, longitude, and time of day.

    a) Pseudoscience
    b) Scientific method
    c) Determinism
    d) Observation

33. Which of the following scale is best representative of a small-scale map?

    a) 1 inch to 1 mile
    b) 1:1,000,000
    c) 1:63,360
    d) 1:24,000

34. A map in which the area of its units has been distorted to be proportional to the data they represent is known as ________

    a) Cartogram
    b) Equal area
    c) Proportional symbol
    d) Topograhic

35. Water pollution, such as the amount of sediment, may be detected using which of the following?

    a) Astrological images
    b) Orthophoto images
    c) Topographic images
    d) Thermal scanners

36. A map scale of 1:63,360 means that ________

    a) One inch on the map represents 63,360 miles on the earth

    b) One inch on the map represents 63,360 inches on the earth

c) One inch on the map represents 63,360 feet on the earth

d) One foot on the map represents 63,360 inches on the earth

37. The term 'remote sensing' refers to the ________

a) Calculation of the precise location of all one million plus U.S.G.S. benchmarks

b) Construction of maps by computers

c) Detection of the nature of objects from a distance

d) Wavelengths of the electromagnetic spectrum

38. General purpose maps that depict the shape and elevation of terrain, and usually portray the surface features of relatively small areas, are known as ________

a) Topographic maps

b) Thematic maps

c) Mercator maps

d) Value-by-area maps

39. The distance between each degree of latitude is approximately ________

a) 65 km

b) 100 km

c) 111 km (69 miles)

d) 360 km

i) 33 miles

ii) 50 miles

iii) 69 miles

iv) 111 miles

40. Which of the following properties of a globe is incorrect?

a) All lines of latitude are parallel to the equator and to each other.

b) All meridians converge at the poles and are true north-south lines

c) Meridians and parallels intersect at right angles

d) Parallels increase in length as one nears the poles.

41. A half circle of 180 degrees of arc which connects the earth's poles is known as a ________

a) Latitude

b) Meridian

c) Longitude

d) Parallel

42. The system of angular measurement for specifying a location north or south of the equator is known as ________

    a) Latitude   b) Longitude

    c) Parallel   d) Meridian

43. Which statement is true about GPS technology?

    a) GPS uses remote sensing

    b) GPS is a navigational device

    c) GPS involves infrared photo

    d) GPS gives relative location of places

44. To increase precision, latitude and longitude are divided into ________

    a) Metres and centimetres   b) Minutes and seconds

    c) Latimites and longimites   d) Meridians

43. The contour interval represents the ________

    a) Accuracy of the surveying work

    b) Horizontal spacing between the contour line

    c) Number of contour lines on a m

    d) Vertical spacing between contour lines

44. Which of the following is not a map projection?

    a) Equidistant   b) Geoidal

    c) Conformal   d) Azimuthal

45. Each time zone extends across approximately how many degrees of longitude?

    a) 15   b) 50

    c) 69   d) 111

46. Geographic information systems ________

    a) Digitizes data from many different sources in order to display any combination of variables for map analysis.

    b) Are limited to military and homeland defense applications

    c) Emphasize the use of the map as a means of information storage

    d) Can only display information from government data

47. The principal purpose of topographic maps is to ________
    a) Assess the human impact upon environmentally sensitive areas
    b) Determine the location and extent of urban areas
    c) Display elevation and terrain by the use of contours
    d) Provide information about geologic processes

48. If a map is called conformal, then it is correct for areas in terms of ________
    a) Area b) Shape
    c) Distance d) Direction

49. A large-scale map depicts ________
    a) Large amount of detail for a large amount of area
    b) A large amount of detail for a small amount of area
    c) A small amount of detail for a large amount of area
    d) A small amount of detail for a small amount of area

50. What is the principal device to show elevation on a topographic map?
    a) Benchmarks b) Spot heights
    c) Shaded relief d) Contour lines

51. The map type best used to record not only the presence of a phenomenon but also to suggest a visual impression of the pattern is ________
    a) Dot b) Choropleth
    c) Linear d) Cartogram

52. Flowlines would not be used to represent ________
    a) Airline routes b) Political boundaries
    c) Migration patterns d) Ocean currents

53. False colored maps are produced by ________
    a) Thermal scanners b) Radar
    c) Infrared film d) Standard photographic film

54. A choropleth map and a cartogram are both examples of the portrayal of geographic information based on ________
    a) Point symbols
    b) National symbols
    c) Line symbols
    d) Area symbols
55. The Global Positioning System is maintained by the ________
    a) United Nations
    b) US Department of Defense
    c) US Geological Survey
    d) Automotive industry
56. The Prime Meridian runs through ________
    a) New York City, New York
    b) Quito, Ecuador
    c) Greenwich, England
    d) Pacific Ocean
57. Which of the following are correct latitude and longitude coordinates?
    a) 110 degrees North, 78 degrees East
    b) 110 degrees East, 78 degrees North
    c) 45 degrees South, 120 degrees East
    d) 60 degrees South, 123 degrees North
58. Which type of map projection shows true directions from one central point to all other points?
    a) Equivalent
    b) Conformal
    c) Equidistant
    d) Azimuthal
59. What kind of map shows a specific spatial distribution or category of data?
    a) Thematic map
    b) Reference map
    c) General-purpose map
    d) Location map
60. What kind of map combines the image characteristics of a photography with the geometric characteristics of a map?
    a) Aerial topographic map
    b) Electromagnetic map
    c) Stereoscopic map
    d) Orthophoto map
61. Which statement is not correct regarding Landsat satellites?
    a) The first Landsat satellite was launched in 1972
    b) Landsat satellites are in a polar orbit from south to north
    c) Landsat's sensors measure many different types of electromagnetic wavelengths.
    d) Landsat's sensors can depict objects upto 15 metres (50 feet) below the ground.

62. What term refers to the art, science and technology of making maps?
    a) Geography
    b) Geographical Information Systems
    c) Cartography
    d) Topography
63. Which of the following representative fractions would show the most detail on a map?
    a) 1:1,000,000 b) 1:24,000
    c) 1:63,360 d) 1:10,000,000
64. What type of map symbol do contour lines represent?
    a) Isoline b) Point
    c) Area d) Flow line
65. The statement "1 centimetre to 1 kilometre" is an example of what type of scale?
    a) Metric b) Verbal
    c) Graphic d) Representative fraction
66. Deforestation is occurring in many countries, and it is, therefore, difficult to detect the ongoing damage. Which of the following types of remote sensing would be best suited for locating deforestation?
    a) Thermal infrared b) Microwave
    c) Radar d) Sonar
    e) Colour infrared
67. Multi-spectral remote sensing allows researchers to obtain data about the Earth in areas of the electromagnetic spectrum beyond what the eye can see. If you were studying the effects of drought on vegetation, which portion of the electromagnetic spectrum would provide the most pertinent data?
    a) Thermal infrared b) Middle infrared
    c) Near infrared d) Blue spectrum
    e) Red spectrum

68. Which of the following types of remote sensing would be most useful in obtaining an accurate terrain representation of Venus?

a) Radar
b) Microwave
c) Landsat
d) Sonar
e) Thermal infrared

69. With which type of remote sensing imagery would a baseball field of artificial turf be differentiated from natural grass?

a) Radar imagery
b) Color infrared
c) Black and white photography
d) Color photography
e) All the above

70. Which of the following remote sensing technologies uses sound?

a) Radar
b) Color infrared imaging
c) Thermal infrared imaging
d) Sonar

71. The first step in developing a GIS is to create a digital record of geographic information known as a ________

a) Digital depository
b) Geographic database
c) Computerized repository
d) Geofile

72. A computer-based set of procedures for handling geographically referenced information is known as ________

a) Global positioning systems (GPS)
b) Geographic information systems (GIS)
c) Digitally integrated geography (DIG)
d) Intelligent geographic design (IGD)

73. A relatively new remote sensing technology that utilizes an airborne laser to transmit light out to an object is known as ________

a) Thermal scanning
b) Radar
c) Lidar
d) Global positioning systems

74. A map maker is called a _______
   a) Photographer b) Cartographer
   c) Map quester d) Geographer

75. Remote Sensing is _______
   a) Deployed from a computer
   b) Deployed from a plane or satellite
   c) Always done close up
   d) Deployed from a camera and video camera

78. The First type of remote sensing were _______
   a) TV remote controls
   b) Sketches from a French Map Maker
   c) Paintings from an air balloon
   d) Aerial photos

79. Sensing that uses Heat or Temperature to detect is called ________
   a) Microwave Sensing b) Radar Sensing
   c) Infrared Sensing d) Sonar Sensing

80. Infrared sensing is used to help map _______
   a) Animals that have died b) Animals and plants
   c) Non-living objects d) Cold blooded organisms

81. Microwave Sensing uses _______
   a) Heat and temperature b) Very long wave lengths
   c) Short wave lengths d) Sound waves

82. Which two sensing methods bounce waves off of an object and back to detect things?
   a) Radar and sonar b) Radar and infrared
   c) Touch and tasting d) Microwave and sonar

83. Sound waves are used in what type of sensing?
   a) Microwave b) Infrared
   c) Sonar d) Radar

84. What sensing is used under water to detect things

    a) Sonar
    b) Microwave
    c) Infrared
    d) Radar

85. The French used what two items to create maps in the 1800's?

    a) Planes and Satellites
    b) Hot Air Balloon and a Primitive Camera
    c) A Parachute and an Artist to Draw what he saw
    d) A Spy Plane and a 35mm Camera

86. Remote Sensing can be defined as ________

    a) A TV remote to change channels
    b) Collecting information through a remote controlled device
    c) Collecting information from a distance by photos or wave lengths
    d) Collecting information by a person on two legs and a camera

87. When we see the colours in the rainbow, we are actually seeing ________

    a) Every kind of light there is
    b) A man named Roy G Biv
    c) A small portion of the electromagnetic spectrum
    d) The entire Electromagnetic Spectrum

88. Infrared Sensing is used commonly to study ________

    a) Crops and different habitats
    b) Objects that can only be picked up using sound waves
    c) Underwater sealife
    d) Very cold and distant objects

89. Radar is very beneficial because it can penetrate thick ________

    a) Sand and very dry objects
    b) Walls and concrete
    c) Clouds and moisture
    d) Steel and wood

90. Remote Sensing is NOT good to use when comparing an area before a Tsunami and then after the Tsunami ________
    a) Remote Sensing has nothing to do with Tsunamis
    b) False
    c) Could be either answer
    d) True

91. The ratio of reflected energy to incident energy as a function of wavelength is ________
    a) Spectral reflectance b) Irradiance
    c) Diffusion d) Conductance

92. Spectral reflections are undesirable as they give rise to __________ on aerial photographs.
    a) Scattering b) Hotspots
    c) Distortions d) Aberations

93. Suspended sediments in a water body _____________ the spectral reflectance.
    a) Increase b) Decrease
    c) Remains same d) Reflects

94. Best season for taking aerial photography of teak is ________
    a) April–May b) June–July
    c) Sept-Oct d) Feb-Mar

95. The intensity of EMR reflected from a surface depends ________
    a) Surface roughness b) Wavelength
    c) Polarization d) All

96. Which one of the following has low reflectance?
    a) Vegetation b) Soil
    c) Water d) Air

97. The relationship between scale factor S, flying height H and lens focal length f is ______
    a) s=fH b) s=f/H
    c) s=H/f d) s=1/fh

98. X, Y coordinate used to describe a geographic feature is ________

    a) Raster data b) Vector data

    c) Topology d) None of the above

99. The techniques of making measurements from photos or image data is called ________

    a) Photogrammetry b) Cartography

    c) Digital Image Processing d) GIS

100. Parameter that describes the electrical properties of a medium is ____________.

    a) Emissivity b) Dielectric constant

    c) Latent heat d) Radiation

101. The relationship between the radiation emitted from blackbody and its temperature is explained by ________

    a) Kirchoff's law b) Wien's displacement law

    c) Stefan - Boltzman's law d) Planck's law

102. ____________________ influences a broad range of wavelengths in and near visible spectrum.

    a) Mie scattering b) Rayleigh scattering

    c) Non-selective scattering d) Both a and b

103. Survey of India Topo - sheet is a ________

    a) Perspective projection b) Orthogonal projection

    c) Polyconic projection d) Universal projection

104. Sun-synchronous satellites travel from ________

    a) S-N b) N-S

    c) E-W d) W-E

105. The ability of sensor to discriminate very slight differences in energy is called ________

    a) Spectral resolution b) Radiometric resolution

    c) Temporal resolution d) Spatial resolution

106. SAR (Synthetic Aperture Radar) is an example of ________

a) Active sensor
b) Passive sensor
c) Microwave sensor
d) None

107. Among the following sensor which has the highest spatial resolution?

a) TM
b) ETM
c) WIF's
d) LISS

108. Microwave remote sensing utilizes ________

a) 0.3–3.0 mm
b) 1mm–1m
c) 3-5 mm
d) 1-5 mm

109. High oblique photographs have tilt of ________

a) $> 40^{o}$
b) $> 60^{o}$
c) $> 30^{o}$
d) $< 40^{o}$

110. The best season for taking aerial photographs of teak forests of south India is ________

a) March–April
b) September–December
c) October–November
d) September–October

111. Ground receiving station in India is located at ________

a) Hassan
b) Shadnagar
c) Ahmedabad
d) Bangalore

112. Scanning system which collects the data over range of wavelengths is called as ________

a) MSS
b) WIFS
c) TM
d) None

113. Which of the following wavelength band is used for vegetational studies?

a) UV
b) Visible only
c) Visible & IR
d) IR, Visible & UV

114. Which of the following remote sensing satellite was launched first by India?

a) IRS-IC
b) IRS-IA
c) IRS-1
d) IRS-P1

115. Relative brightness of objects in an image is referred as ________

a) Texture
b) Tone
c) Pattern
d) None

116. Which of the following recently launched satellite having 1m resolution?

a) IRS-IE
b) LANDSAT-7
c) IKONOS
d) SPOT-III

117. GIS uses ________

a) Spatial data
b) Non spatial data
c) Attribute data
d) Both a & b

118. Which of the following is very useful in detecting and mapping forest fires?

a) Visible sensors
b) Thermal sensors
c) Microwave sensors
d) None

119. Aerial photograph taken with optical axis of approximately perpendicular to the horizontal plane and has deviation upto 4 degrees ________

a) Oblique
b) High oblique
c) Vertical
d) Low oblique

120. Remote sensing and GIS technology can be applied for ________

a) Biodiversity characterization
b) Vegetation classification
c) Wildlife management
d) All of these

Q II. Match the Following

Set I

S.No	Part A	Part B
1.	Attribute data	Frequency of tonal change
2.	Passive sensor	EMR from rough surface
3.	NRSC	Parallax
4.	Push broom scanner	Tables, Texts
5.	Texture	Hyderabad
6.	Pattern	Ground resolution
7.	Spatial Resolution	Across the track
8.	Diffused reflection	Sun
9.	Whiskbroom Scanner	Spatial arrangement of objects
10.	Stereo plotting	Charged coupled devices

Set II

S.No	Part A	Part B
1.	Optical remote sensing system	Eyes in the sky
2..	Visual remote sensing system	Penetrate thick clouds and moisture
3.	Airborne Remote Sensing	Images acquired in day and night
4.	Radar Remote Sensing	Spectral reflectance signatures
5.	Acoustic and near-acoustic remote sensing	High spatial resolution images
6.	Microwave Remote Sensing	Detection of forest fires, volcanoes and oil fires
7.	Satellite Remote Sensing	Sonar and Seismograms
8.	Longest-wavelength portion of the EMS	Passive Remote Sensing
9.	Cartography	Radio and TV Waves
10.	Infrared Remote Sensing	Map and chart construction

Q III. State True or False

1. Spectral character of the vegetation does not vary with wavelength.
2. Small scale photographs have poor resolution.
3. Presence of atmospheric windows is not desirable for RS.

4. TIROS is an example for thermal satellite
5. Ideal reflection for remote sensing is specular reflection.
6. Majoring and monitoring UV, NIR, MIR is one way to study the vegetation condition.
7. If scale is fixed, focal length does not vary with the flying height of the aircraft.
8. Texture is resultant of differential reflection of light.
9. Irradiation is the time rate of flow of energy.
10. Geostationary orbits are low altitude orbits.
11. Digital number is a value assigned to a pixel in a digital image

Q IV. Fill in the Blanks

1. Theory of Electromagnetic Spectrum was given by ___________
2. Two types of remote sensing .___________and ___________.
3. RADAR is an example of ___________ remote sensing (type).
4. Lillesand and Keifer postulated that ___________ is the basis on which whole remote sensing technique is based.
5. In India ___________ is the Govt. agency that carry aerial photography.
6. Expand NDVI ___________.
7. The technique of acquiring quantitative information from aerial photographs is termed as ___________
8. Emissivity of an object is explained by ___________ law.
9. Energy passing through a surface in a unit time is referred as ___________
10. For large scale photography, ___________ focal length is desirable on flat and gently undulating areas.
11. Expand NRSC ___________ and is located in ___________
12. ______ is the instrument used for the study of aerial photographs.
13. LANDSAT satellite mission belongs to ___________ country.
14. Expand NRSC___________ and it is located in ___________.

15. Manipulation of the digital - number values of an image by a Computer is known as Digital image processing.

16. Expand LASER - Light artificially stimulated electromagnetic radiation.

17. Planck's Law states an expression for the variation of emittance of a blackbody at a particular temperature as a function of wavelength.

18. Expand Radar - Radio detection and ranging.

19. The sensors in the human visual system are the two types of photosensitive cells, known as the ________ and the __________, at the retina of the eyes.

20. The human visual system is an example of a ____________ Remote Sensing system which needs an external source of energy to operate.

21. Stereoscope is a binocular optical device for viewing overlapping images or diagrams in which the left eye sees only the left image, and the right eye sees only the right image.

22. Camera systems are passive optical sensors that use a lens or system of lenses (optics) to form an image at the ______________ at which an image is sharply defined.

23. Panchromatic film produces ____________ and ____________ images and is the most common type of film used for aerial photography.

24. In a false colour photograph, targets with high near-infrared reflectance appear _____________, those with a high red reflectance appear _____________, and those with a high green reflectance appear ___________, thus giving us a "false" presentation of the targets relative to the colour.

25. Expand KSRSAC __________________________and Govt. of Karnataka established KSRSAC in the year ______________.

26. Theory of ________________________________was given by J.C. Maxwell in 1873.

27. _____________________ is an example of active remote sensing where the time delay between emission and return is measured, establishing the location, height, speeds and direction of an object.

28. The Word "photography" literally means ________________.

29. ________________ is a dynamic form of energy that propagates as wave motion at a velocity of light.

30. In an interaction of EMR with the earth's surface, $E_I(g)= E_R(g)+E_A(g)+E_T(g)$ where $E_I(g)=$________________ $_R(g)=$__________ $E_A(\gamma)=$____________________ $E_T(\gamma)=$______________________.

31. The components of GIS i) ____________ ii) ____________ iii) ____________ iv) ____________ v) ____________.

32. ________________, ____________ and ________________ are the primary colours or wavelengths of the visible spectrum. (BGR)

33. Expand LIDAR - Light intensity detection and ranging

34. Path of a satellite around a body, such as the earth, under the influence of gravity is called as orbit.

35. Reflectance is the ratio of the radiant energy reflected by a body to the energy incident on it.

36. When Radio waves travel at the speed of light, it takes 0.06 seconds to receive a signal transmitted by a satellite floating directly overhead. Then the distance from the satellite is ________________ kms.

37. The Word "photography" literally means ________________.

38. Kirchhoff's Law is the radiation law which states that at a given temperature the ratio of the emissivity to the absorptivity for a given wavelength is the same for all bodies and is equal to the emissivity of an ideal blackbody at that temperature and wavelength.

39. Wien's displacement law explains the shift of the radiant power peak to shorter wavelengths as temperature increases.

40. Aerial photography is synonymously called as ______________.

41. A set of powerful tools for collecting, storing, retrieving at will, transforming and displaying spatial data from the real world, for a set of purposes is known as ____________________.

42. A science of collecting and interpreting information about a target without being in physical contact with the object is referred as ________________________________.

43. Remote sensing deals with the collection and interpretation of information about an object without being in physical contact with the object.

44. Write three Segments of the GPS i) ________ ii) __________ iii) ____________________.

45. GPS with four or more satellites in view, the receiver can determine the user's 3D position. i) _____________ ii) ______________ iii) _______________________.

46. Write the name of the scientist who took an oblique photograph of a small village near Paris from a balloon in 1859 : ________ ________________________________.

47. Expand Sonar - Sound navigation ranging

48. Functional subsystems of A GIS i) ____________________ ii) _________________________ iii) ________________________ iv) __________________.

49. Types of Aerial photographs based on direction of exposure i) ________________________ ii) ________________________.

Q V. Expand the following abbreviations in relation to remote sensing and GIS

1.	EMS	2.	CIR
3.	CPU	4.	GPS
5.	GIS	6.	VNIR
7.	SWIR	8.	WIR
9.	LWIR	10.	SAR
11.	IIRS	12.	NRSA
13.	ISRO	14.	SDSC
15.	LPSC	16.	NESAC
17.	GSLV	18.	VSSC
19.	PSLV	20.	EOS
21.	SAC	22.	TIROS
23.	KSRSAC	24.	NASA
25.	NRSC	26.	ISRO
27.	ERTS	28.	LIDAR
29.	GMT	30.	SONAR
31.	RADAR	32.	LASER

Q VI. Define the Following in one or two Sentences

Remote sensing	Global Positioning Systems	Aerial photograph
Resolution	EMR	Panchromatic image
Photogrammatry	Normal colour photograph	Camera
Filtering	GIS	Oblique aerial photographs
Drift	Accuracy	Vertical photographs
Passive remote sensing	Active remote sensing	Photo Interpretation
Photons	EMS	Spectral Signature
Rayleigh Scattering	Mie Scattering	Non-selective Scattering
Oblique aerial photographs	Vertical photographs	Multiband photography
Black body	Focal length	Stereoscopy
Camera	False colour photograph	Photogrammetry
Geoinformatics	Atmospheric Windows	Digital number
Atmospheric Absorption	Digital Image	Mie Scattering

Q VII. Answer the Following / Write the Short Notes

1. Global Positioning System (GPS)
2. Airborne platforms
3. Stages of remote sensing
4. Modes of scanning
5. Spatial and Spectral resolution
6. Advantages of Remote Sensing
7. Elements of photo interoperation
8. Applications of Remote Sensing.
9. List the types of remote sensing systems.
10. Atmospheric Scattering
11. Write four Laws of Light and explain briefly
12. Write the applications of Cameras
13. Briefly explain Optical Remote Sensing system and Infrared Remote Sensing system
14. Main types of remote sensing
15. Write about the basic concept of GPS.
16. Short notes on Electromagnetic radiation and the electromagnetic spectrum
17. Write a note on Reflectance Characteristics of the Earth's Cover types.

Q VIII. Answer the Following

1. What is EMS? Explain with help of neat diagram.
2. Explain reflectance bahaviour of natural vegetation with help of neat diagram.
3. What is photogrammatry? Explain tree height measurement on aerial photography using relief displacement method.
4. What are the different applications of remote sensing techniques?
5. Define geoinformatics and what are its applications?
6. What is DIP? Mention different steps involved in DIP and explain classification.
7. Explain the principles and process of remote sensing with a neat diagram.
8. Explain the interaction of EMR with the earth's surface with neat diagram.
9. Explain some characteristics of Black body radiation
10. Write a short note on Oblique aerial photographs and Vertical photographs.
11. Explain photogrammetry and its classes. Write the applications of photogrammetry.
12. Explain any five types of remote sensing systems with neat diagrams
13. Explain the principles and process of remote sensing with neat diagram.

❑❑❑

Chapter 14

Appendices

Appendix - I

List of Common Physics Notations

This is a list of common physical constants and variables & their notations. Note that bold text indicates that the quantity is a vector.

Latin characters

Symbol	Meaning	SI unit of measure
A	area	meter squared (m^2)
	magnetic vector potential	
	amplitude	
a	acceleration	meter per second squared (m/s^2) or ($m\ s^{-2}$)
B	magnetic flux density also called the magnetic field density or magnetic induction	tesla (T), or equivalently, weber per square metre (Wb/m^2)
C	capacitance	farad (F)
	heat capacity	joule per kelvin ($J\ K^{-1}$), or equivalently, joule per degree Celsius ($J\ °C^{-1}$)
	constant of integration	varied depending on context
c	speed of light (in vacuum)	299,792,458 metre per second (m/s)
	speed of sound	340.29 meter per second (m/s)
	specific heat capacity	joule per kilogram per kelvin ($J\ kg^{-1}\ K^{-1}$), or equivalently, joule per kilogram per degree Celsius ($J\ kg^{-1}\ °C^{-1}$)
	viscous damping coefficient	kilogram per second (kg/s)
D	electric displacement field also called the electric flux density	coulomb per square metre (C/m^2)
d	distance	meter (m)
	impact parameter	meter (m)
	diameter	meter (m)
	differential (e.g. dx)	
$d\mathbf{A}$	differential vector element of surface area A, with infinitesimally small magnitude and direction normal to surface S	square meter (m^2)
dV	differential element of volume V enclosed by surface S	cubic meter (m^3)
E	electric field	newton per coulomb ($N\ C^{-1}$), or equivalently, volt per meter ($V\ m^{-1}$)

E	energy	joule (J)
e	eccentricity	unitless
	2.71828... (base of the natural logarithm), electron, elementary charge	
$\mathbf{F}$	force	newton (N)
f	frequency	hertz (Hz)
	function	
	friction	newton (N)
G	the gravitational constant	newton metre squared per kilogram squared ($N\ m^2/kg^2$)
g	acceleration due to gravity	meter per second squared (m/s^2)
$\mathbf{H}$	magnetic field strength also called just magnetic field	ampere per metre (A/m)
H	Hamiltonian	joule (J)
h	height	meter (m)
	Planck's constant	joule second (J s)
$\hbar$	reduced Planck's constant $\left(\frac{h}{2\pi}\right)$	joule second (J s)
I	action	joule second (J s)
	intensity	watt per square metre (W/m^2)
	sound intensity	watt per square metre (W/m^2)
	electric current	ampere (A)
	moment of inertia	kilogram meter squared ($kg\ m^2$)
i	intensity	watt per square metre (W/m^2)
	imaginary unit	
$\hat{\mathbf{i}}$	Cartesian x-axis basis unit vector	unitless
$\mathbf{J}$	*free* current density, not including polarization or magnetization currents bound in a material	ampere per square metre (A/m^2)
	impulse	kilogram meter per second (kg m/s)
$\hat{\mathbf{j}}$	Cartesian y-axis basis unit vector	unitless
K	kinetic energy	joule (J)
k	Boltzmann constant	joule per kelvin (J/K)
	wavenumber	radians per meter (m^{-1})
$\hat{\mathbf{k}}$	Cartesian z-axis basis unit vector	unitless
L	inductance	henry (H)
	luminosity	watt (W)
	angular momentum	newton metre second (N m s or $kg\ m^2\ s^{-1}$)

l	length	metre (m)
M	magnetization	ampere per metre (A/m)
	moment of force often simply called moment or torque	newton metre (N m)
m	mass	kilogram (kg)
N	normal vector	unit varies depending on context
	atomic number	unitless
n	refractive index	unitless
	principal quantum number	unitless
P	power	watt (W)
p	momentum	kilogram metre per second (kg m/s)
	pressure	pascal (Pa)
Q	electric charge	coulomb (C)
	Heat	joule (J)
q	electric charge	coulomb (C)
R	electrical resistance	ohm (Ω)
	Ricci tensor	unitless
	radiancy	
r	radius vector (position)	metre (m)
r	radius of rotation or distance between two things such as the masses in Newton's law of universal gravitation	meter (m)
S	surface area	m^2
	entropy	joule per kelvin (J/K)
	action	
s	arc length	metre (m)
	displacement	
T	period	second (s)
	thermodynamic temperature also called absolute temperature	kelvin (K)
t	time	second (s)
U	four-velocity	meter per second (m/s)
U	potential energy	joule (J)
	internal energy	joule (J)
u	relativistic mass	kilogram (kg)

	energy density	joule per cubic metre (J/m^3) or joule per kilogram (J/kg) depending on the context
V	voltage also called electric potential difference	volt (V)
	volume	cubic metre (m^3)
	shear force	
v	velocity	meter per second (m/s)
W	mechanical work	joule (J)
w	width	meter (m)
x	a generic unknown	varied depending on context
	displacement	meter (m)
Z	electrical impedance	ohm (Ω)

Greek characters

Symbol	Name	Meaning	SI unit of measure
α	alpha	angular acceleration	radian per second squared (rad/s^2)
β	beta	velocity in terms of the speed of light c	unitless
γ	gamma	Lorentz factor	unitless
		photon	
		gamma ray	
		shear strain	
		Heat capacity ratio	unitless
Δ	delta	a change in a variable (e.g. Δx)	unitless
		Laplace operator	
δ	delta	displacement (usually small)	
ϵ	epsilon	permittivity	farad per meter (F/m)
		strain	unitless
ζ	zeta	damping ratio	unitless
η	eta	energy efficiency	unitless
		coefficient of viscosity also called simply viscosity	pascal second (Pa s)
θ	theta	angular displacement	radian (rad)
K	kappa	torsion coefficient also called torsion constant	newton meter per radian (N m/rad)
Λ	lambda	cosmological constant	per second squared (s^{-2})
		wavelength	metre (m)

μ	mu	magnetic moment	ampere square metre ($A\ m^2$)
		coefficient of friction	unitless
		dynamic viscosity	Pascal second (Pa s)
		permeability (electromagnetism)	Henry per metre (H/m)
ν	nu	frequency	hertz (Hz)
		kinematic viscosity	meters squared per second (m^2/s)
π	pi	3.14159... (irrational number)	
ρ	rho	mass density usually simply called density	kilogram per cubic metre (kg/m^3)
		free electric charge density, not including dipole charges bound in a material	coulomb per cubic metre (C/m^3)
		resistivity	Ohm metre (Ω)
$\sum$	sigma	summation operator	
σ	sigma	electrical conductivity	Siemens per meter (S/m)
		normal stress	
τ	tau	torque	newton metre (N m)
		shear stress	
		time constant	second (s)
Φ	phi	field strength	unit varies depending on context
		magnetic flux	Weber (Wb)
ϕ	phi	electric potential Higgs Field	
Ψ	psi	wave function	unitless
ω	omega	angular frequency	radian per second (rad/s)

Other characters

Symbol	Name	Meaning	SI unit of measure
$\nabla\cdot$	nabla dot	the divergence operator often pronounced "del dot"	per metre (m^{-1})
$\nabla\times$	nabla cross	the curl operator often pronounced "del cross"	per metre (m^{-1})
∇	nabla	del (differential operator)	
∂	"der", "dow", "die", "partial" or simply "d"	partial derivative (e.g. $\partial y/\partial x$)	

Appendix II

Forest Inventory

Forest inventory is the systematic collection of data and forest information for assessment or analysis. An estimate of the value and possible uses of timber is an important part of the broader information required to sustain ecosystems.[1] When taking forest inventory, the following are important things to measure and note: species, diameter at breast height (DBH), height, site quality, age, and defects. From the data collected, one can calculate the number of trees per acre, the basal area, the volume of trees in an area, and the value of the timber. Inventories can be done for other reasons than just calculating the value. A forest can be cruised to visually assess timber and determine potential fire hazards and the risk of fire. The results of this type of inventory can be used in preventative actions and also awareness. Wildlife surveys can be undertaken in conjunction with timber inventory to determine the number and type of wildlife within a forest. The aim of the statistical forest inventory is to provide comprehensive information about the state and dynamics of forests for strategic and management planning.

History

Surveying and taking inventory of trees originated in Europe in the late 18th century out of a fear that wood (the main source of fuel) would run out. The first information was organized into maps used to plan out usage. In the early 19th century forest harvesters estimated the volume and dispersal of trees within smaller forests with their eyes. More diverse and larger forests were divided into smaller sections of similar type trees that were individually estimated by visual inspection. These estimates were related together to figure out the entire forest's available resources. As the 19th century progressed so did the measurement techniques. New relationships between diameter, height, and volume were discovered and exploited. These newfound relationships allowed for a more accurate assessment of wood types and yields of much larger forests. By 1891, these surveys were conducted through sample-based methods involving statistical averages and more sophisticated measuring devices were implemented. In the 20th century, the statistical method of sampling had become well established and commonly used. Further developments, such as unequal probability sampling, arose. As the

20th century progressed, an understanding of coefficients of error became clearer and the new technology of computers combined with the availability of aerial as well as satellite photography, further refined the process. Laser Scanning both terrestrially and aerially are now used alongside more manual methods. As a result sampling accuracy and assessment values became more accurate and allowed for modern practices to arise.

A forest inventory does not only record the trees height, DBH and number for tree yield calculations. But it also records the conditions of the forest. So, this would include geology, site conditions, tree health and other forest factors. Merely looking at the forest for assessment is called taxation.

Timber cruise

A timber cruise is a sample measurement of a stand used to estimate the amount of standing timber that the forest contains. These measurements are collected at sample locations called plots or quadrants. Each of these individual plots is one observation in a series of observations called a sample. These plots are generally laid out in some random fashion usually in the form of a line plot survey. Depending on the size of the plot and the number of plots measured, the data gathered from these plots can then be manipulated to achieve varying levels of certainty for an estimate that can be applied to the entire timber stand. This estimate of stand conditions, species composition, volume and other measured attributes of a forest system can then be used for various purposes. For example, in British Columbia the sale of Crown timber is a business proposition and both the buyer and the Ministry of Forests and Range (seller) must know the quantity and the quality of timber being sold. The cruise provides the essential data for determining stumpage rates, for establishing conditions of sale and for planning of the logging operations by the licensee.

Stand Examination

Example of hardware equipment for forest inventories: GPS and laser rangefinder for mapping connected to a field rugged computer.

Today, the most common type of inventory is one that uses a random sampling technique which groups similar forests into one category based on age, stand structure, species, and location. The next step would be to begin from a random place and measure circular plots that are equidistant from one another. Inventory undertaken in conjunction with wildlife surveys has a regularized distance between plots. When taking inventory, a decision has to be made about which types of plots to measure. There are several different types of plots. The two most common types are fixed radius and variable radius, also known as prism radius. In a fixed radius plot, the forester finds the centre of a plot and every tree within a certain fixed distance away from that point is measured. Variable radius plots are used more for inventory of volume. During this method, an angle is created and projected from the plot centre and all trees that are larger than the projected angle are measured.

Types of sample plots

Fixed area plot

Fixed area plot sample measurements are taken so that they are a fraction of the entire timber stand. This means that the numbers are all proportional to the actual stand values and that by multiplying by the correct corresponding value, you can obtain the actual tract values. These plots are taken randomly so that each sample point has an equal probability of being included in the random sample. Commonly a $100m^2$ plot is taken.

Variable size plots

A variable size plot is more dependent of the size of the trees. The tract is measured on a series of points and the trees are tallied for being in or out depending on their size and location relevant to the plot center. Usually an angle gauge or a wedge prism are used to gather data for this type of plot. This allows for a very quick estimate of the volume and species of a given tract. A common way of doing this is to select BAF 4.

Transects

Transects are arbitrarily determined lines (to prevent sampling bias) through a stand employed as a linear form of sample plot. They are sometimes referred to as "strip lines."

Plot selection

Plots are samples of the forest being inventoried and so, are selected according to what is looked for.

Simple random sampling

A computer or calculator random number generator is used to assign plots to be sampled. Here random means an equal chance of any plot being selected out of all of the plots available. It does not mean haphazard. Often it is modified to avoid sampling roads, ensure coverage of unsampled areas and for logistics of actually getting to the plots.

Systematic sampling

Commonly this is done by a random point and then laying a grid over a map of the area to be sampled. This grid will have pre-assigned plot areas to be sampled. It means more efficient logistics and removes some of the human bias that may be there with simple random sampling.

Systematic stratified sampling

There is some broad grouping, for example by age classes or soil characteristics or slope elevation. And then plots are chosen from each grouping by another sampling technique. It requires some knowledge of the land first and also trust that the groupings have been done properly. In forestry, it may be done to separate plantation areas from mixed forest for example and reduce the amount of sampling time needed.

Systematic clustered sampling

When it is not possible to make strata for stratified sampling, there may be some knowledge about the forest where it can be said that small groupings are possible. These small groupings of plots, if they are near to each other, form a cluster. These clusters are then randomly sampled with the belief that they are representing the actual mix of the forest. As they are close to each other, there is less waking needed and so, it is more efficient.

Timber metrics

The amount of standing timber that a forest contains is determined from:

- Basal area - defines the area of a given section of land that is occupied by the cross-section of tree trunks and stems at their base

- Diameter at breast height (DBH) - measurement of a tree's girth standardized at 1.3 metres (about 4.5 feet) above the ground.
- Form factor - the shape of the tree, based on recorded trees and commonly then given for calculating tree volumes for a given species. It is usually related to DBH or age class. It is distinct from taper.[2] So, it can be Cone or paraboloid for example.
- Girard form class - an expression of tree taper calculated as the ratio of diameter inside the bark at 16 feet above ground to the to that outside bark at DBH, primary expression of tree form used in the United States
- Quadratic mean diameter - diameter of the tree that coordinates to the stand's basal area
- Site index - a species specific measure of site productivity and management options, reported as the height of dominant and co-dominant trees (site trees)in a stand at a base age, such as 25, 50 and 100 years
- Tree taper - the degree to which a tree's stem or bole decreases in diameter as a function of height above ground. So, it can be sharp or gradual.

Volume estimation

- Stocking - a quantitative measure of the area occupied by trees relative to an optimum or desired level of density.
- Stand Density Index - a measure of the stocking of a stand of trees based on the number of trees per unit area and DBH of the tree of average basal area.
- Volume table - a chart based on volume equations that uses correlations between certain aspects of a tree to estimate the standing volume.
- Stand density management diagram - model that uses current stand density to project future stand composition.

Volume can be calculated from the metrics recorded in a plot sample. For example, if a tree was measured to be 20m tall and with a DBH of 19 cm using previous measured tree data a volume could be approximated according to species. Such a table has been constructed by Josef Pollanschütz in Austria.

Volume of tree = BA X h x f pollanschutz

So, f pollanschütz would be derived from the table and is properly called the Form Factor.

To scale this up to a hectare level the result would have to be multiplied by the number of trees of that size. This is called the blow up factor.

Tools used in inventory

- Biltmore stick - utilizes ocular trigonometry to quickly measure diameter and height

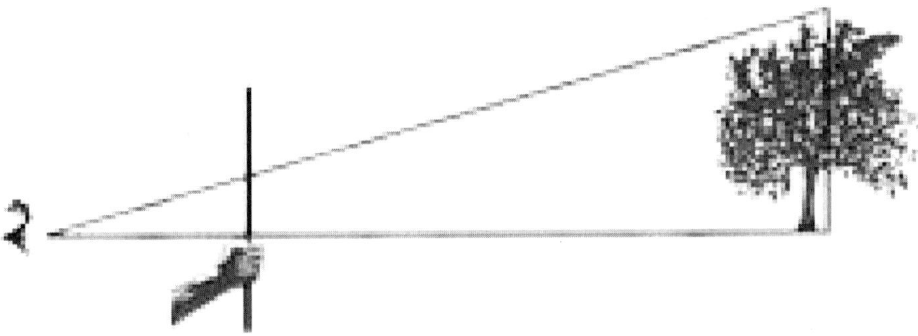

Fig. Demonstrating the ocular trigonometric principles behind the Biltmore stick.

- Diameter tape - cloth or metal tape that is wrapped around the bole, scaled to diameter
- Caliper - two prongs connected to a measuring tape are placed around the most average part of the bole to determine diameter
- Relascope - multiple-use tool that is able to find tree height, basal area, and tree diameter anywhere along the bole
- Clinometer - common tool used to measure changes in elevation and tree height
- Cruising rod - similar to a caliper, calculates the number of pieces of lumber yielded by a given piece of timber by measuring its diameter
- Laser Scanner - used with computer software to calculate the metrics from the collected data by use of Lidar.
- Wedge prism - a small glass wedge that refracts light to create visible offsets in order to be able to choose which trees at a sampling point should be included in the sample.

- Data collector - an electronic device used to quickly enter sample data, geo-locate the data, and, in more modern times, to also access reference, web and historic materials while timber cruising.
- Increment borer - a device used to retrieve a cylindrical sample of wood material orthogonally from the stem while doing as little damage as possible to the remaining tissues.

Appendix III

Estimating Slope

Slope is used by the operations section in several different ways:

- To estimate the amount of time, it takes to construct a fireline;
- To determine whether or not a dozer, engine, or hand crew can work in a specific area;
- To calculate pump pressure needed to reach a location; and
- To calculate fire behaviour characteristics, such as rate of spread.

On incidents, slope is the degree of inclination or steepness and it is usually expressed in per cent. A one per cent slope indicates a rise or drop of one unit over a distance of 100 horizontal units. For example, a one per cent slope rise would indicate a one foot rise over a 100 foot horizontal distance. Slope can be calculated using a topographic map or it can be determined in the field with a clinometer.

To calculate slope using a topographic map, you will need to determine the following:

- **Vertical Distance (also referred to as Rise)** – This is the difference in elevation between two points; it is calculated by subtracting the elevation of one point from the elevation of the other point.
- **Horizontal Distance (also referred to as Run)** – This is the distance from one point to the other and is calculated by measuring distance with a ruler and applying the map scale. For example, if the map scale is 1:24,000 and the distance between the two points when measured with a ruler is ½ inch, the horizontal distance would be 12,000 inches or 1,000 feet.

Slope can then be calculated using the slope formula :

$$\frac{\text{Vertical Distance}}{\text{Horizontal Distance}} \times 100 = \%\,Slope$$

Another way to write the slope formula is: $\frac{\text{Rise}}{\text{Rum}} \times 100 = \%Slope$

There are a number of slope calculation aids in the form of tables that show the relationship between map scale and contour interval. Be careful when using slope calculation aids because they are tailored to specific map scales and contour intervals.

Follow the steps in Table 2-2 to estimate the percent of slope between the two points on the topographic map below :

Table 2-2. Steps to estimate slope between two points.

Steps	Directions
1	Determine the contour interval of the map (40 ft.).
2	Determine the elevation of point A (9480 ft.). Then determine the elevation of point B (9680 ft.).
3	Determine the vertical distance (or rise) between the two points by subtracting the elevation of point A from the elevation of point B (200 ft.).
4	Use a ruler to measure the horizontal distance (or run) between the two points (½ inch). Use the map scale (1:24,000) to convert the ½ inch into feet (1000 ft.).
5	Compute the percentage of the slope using the slope formula: 200 ft. ÷ 1000 ft. = .20 Slope is 20%

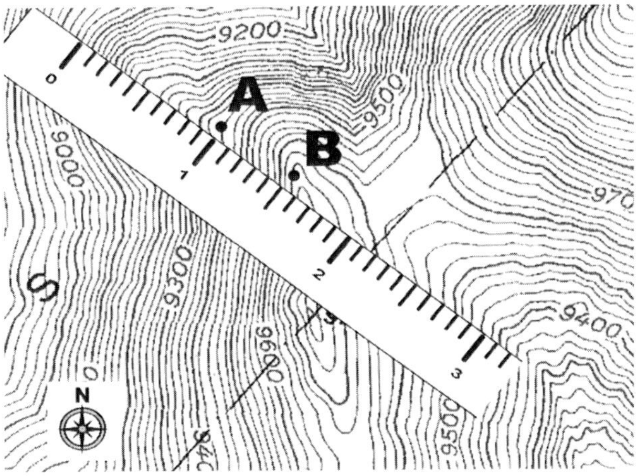

Estimating Aspect

Aspect is the compass direction that the slope is facing. On a topographic map, use index contour lines (they will tell you the difference in elevation) and the north arrow to determine which direction the slope is facing. Follow the steps in Table 2-3 to estimate aspect of the slope between the two points on the topographic map below :

Table 2-3. Steps to estimate aspect

Steps	Directions
1	Determine the elevation of point A (9480 ft.) and point B (9680 ft.).
2	Since point B is a higher elevation than point A, the terrain is sloping from point B down to point A.
3	Use the north arrow on the map to determine which direction the slope is facing, which is northwest. The aspect is northwest.

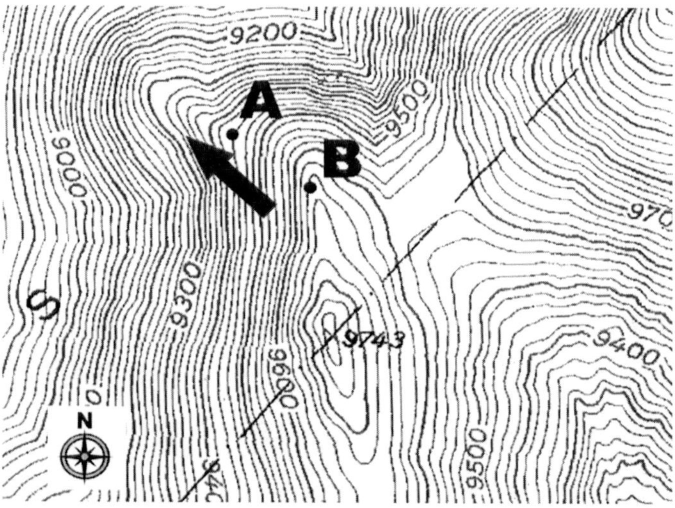

Estimating Acreage

Area can be expressed in square miles, acres, blocks, square feet, or any other square unit of linear measurement. This section discusses different methods for estimating acreage (area formula, dot grid, planimeter, comparison, and GPS receiver). Refer to the Fireline Handbook (PMS 410-1), Appendix A for additional information on estimating acreage.

Area Formula

The most common method for calculating area is using the formula: length x width = area. Area is always calculated in square measure, and the answer will be in square units (square feet, square yards, square chains). When calculating area, typical units of measurements include:

Linear units of measurement	**Area units of measurement**
12 inches = 1 foot	1 acre = 208 feet x 208 feet
3 feet = 1 yard	1 acre = 43,560 square feet
5280 feet = 1 mile	1 acre = 10 square chains
66 feet = 1 chain	640 acres = 1 square mile
80 chains = 1 mile	1 section = 1 square mile*
3.2808 feet = 1 metre	1/2 section = 320 acres*
	1/4 section = 160 acres*
	1 hectare = 2.4 acres

*Generally the size, but may vary due to surveying deviations.

Following are three examples of how to estimate acreage using the area formula:

1. What is the acreage, if the length is 2,640 feet and width is 1,320 feet? Remember that 43,560 square feet equals one acre.

Table 2-4. Estimate acreage using feet.

Steps	Directions
1	Multiply the length by the width to get the area: 2640 ft. x 1320 ft. = 3,484,800 sq. ft.
2	Convert the 3,484,800 sq. ft. to acres: $3{,}484{,}800 \text{ sq. ft.} \times \dfrac{1 \text{ acre}}{43{,}560 \text{ sq. ft.}} = 80 \text{ acres}$

2. What is the acreage, if the length is 5 chains and width is 10 chains? Remember that 10 square chains equals one acre.

Table 2-5. Estimate acreage using chains.

Steps	Directions
1	Multiply the length by the width to get the area: 5 chains x 10 chains = 50 square chains
2	Convert the 50 square chains to acres: 50 square chains x $\frac{\text{1 acre}}{\text{10 chains}}$ = 5 acres

3. What is the acreage for an odd shaped area that is 32 chains wide at one end, 16 chains wide at the other end, and 48 chains in length?

Table 2-6. Estimate acreage for an odd shaped area.

Steps	Directions
1	Multiply the length by the width to get the area: 48 chains x $\frac{16 + 32 \text{ chains}}{2}$ = 1152 square chains Note: To estimate the acreage for an odd shaped area, it is often necessary to take two or more measurements across the area and obtain an average length and width. The average width is obtained by adding the two different widths (16 + 32) and dividing by 2.
2	Convert the 1152 square chains to acres: 1152 square chains x $\frac{\text{1 acre}}{\text{10 chains}}$ = 115.2 acres

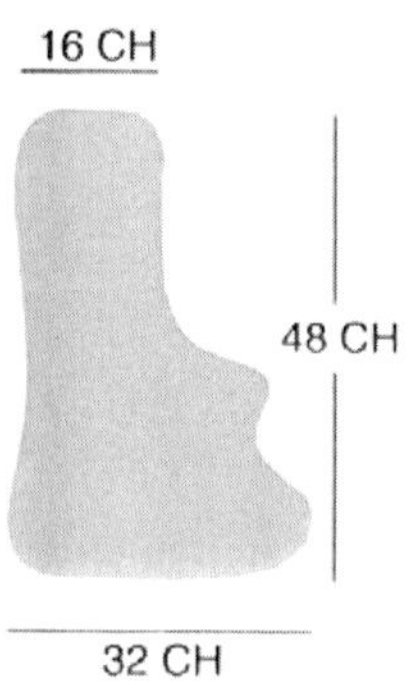

Dot Grid

A dot grid is an inexpensive and readily available tool that is used to estimate acreage (Figure 2-15). There are many different types of dot grids. Each dot represents a specific number of acres depending upon the map scale. For example, if you are using the dot grid in Figure 2-16 on a 7.5 minute quad map, each dot represents 1.434 acres, while on a 15 minute quad map, each dot represents 9.73 acres.

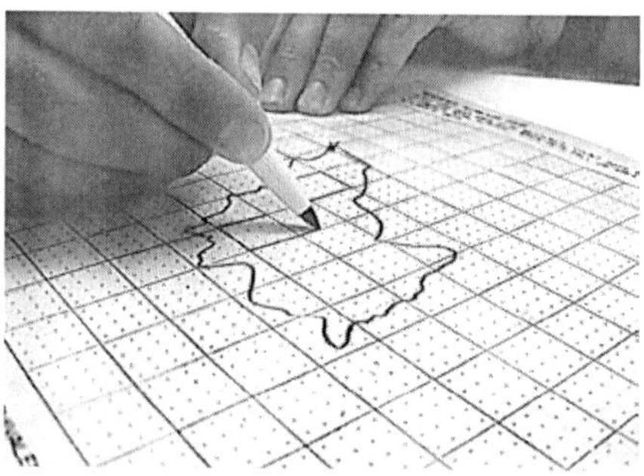

Fig 2-15. A dot grid is used to estimate acreage.

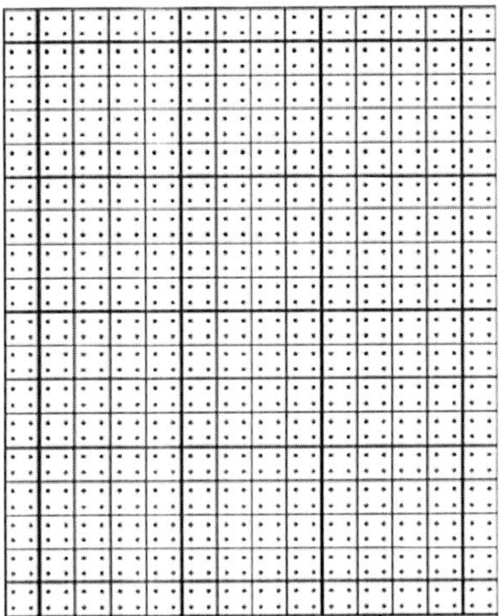

USGS Quadrangle Series	Map Inch Per Mile	Acres Per Square Inch	Each Dot Equals
7.5 Minute Quad	2.64	91.827	1.434 acres
15 Minute Quad	1.01	622.66	9.73 acres

Fig 2-16. This is one example of a dot grid (not to scale).

If a dot grid has no scale or you are working with a map where the scale is questionable, it will be necessary to calibrate the dot grid – which means you need to determine how many acres each dot represents. Table 2-7 describes how to calibrate a dot grid.

Table 2-7. Steps for calibrating a dot grid.

Steps	Directions
1	Determine **map scale.** For this example, we will use a map scale of 1:24000 (1 inch = 2,000 feet).
2	Calculate **acres per square inch.** This will often involve using the conversion: 1 acre = 43,560 square feet. Example: · 1 inch = 2,000 feet 2,000 ft x 2,000 ft = 4,000,000 sq ft $4{,}000{,}000 \text{ sq. ft.} \times \frac{1 \text{ acre}}{43{,}560 \text{ sq. ft.}} = 91.8 \text{ acres per sq. inch}$
3	Determine **acres per dot.** First determine the number of dots per square inch that are on the dot grid. For this example, there are 64 dots per square inch. $\frac{91.8 \text{ acres}}{\text{sq. in.}} \times \frac{\text{sq. in.}}{64 \text{ (dots)}} = 1.434 \text{ acres per dot}$

Table 2-8 describes how to use a dot grid to estimate the acreage of the fire illustrated below :

Table 2-8. Steps for estimating acreage using a dot grid.

Steps	Directions
1	Place the dot grid over the incident perimeter; the position of the dot grid isn't important.
2	Count all the dots that fall completely within the incident perimeter. Record the number.
3	Count every other dot that falls on the boundary of the incident perimeter. Record the number.
4	Add dot counts from steps 2 and 3. This cumulative number of dots represents the total area being measured. Multiply the total count by the area represented by one dot to arrive at the area measurement. In the example below, each dot equals 1.434 acres. 63 (approximate dot count) x 1.434 acres (area represented by one dot) = 90.3 acres

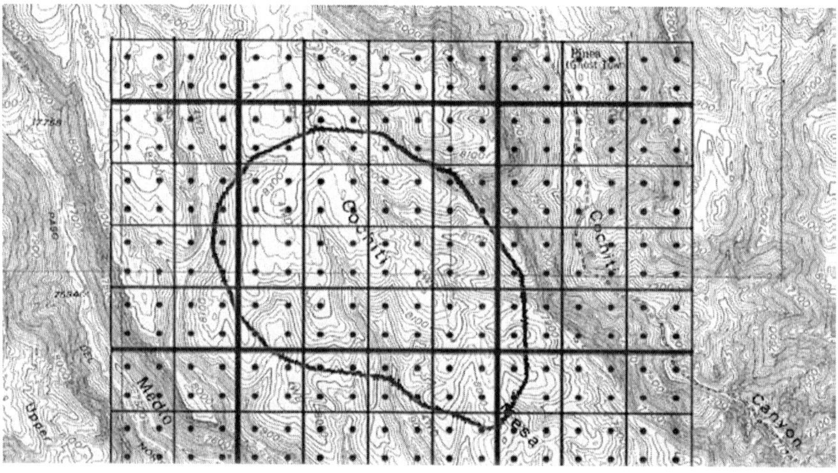

Planimeter

A planimeter is a tool that can be used to measure acreage on a map (Figure 2-17). Use the planimeter to trace around the perimeter a number of times to obtain an average acreage.

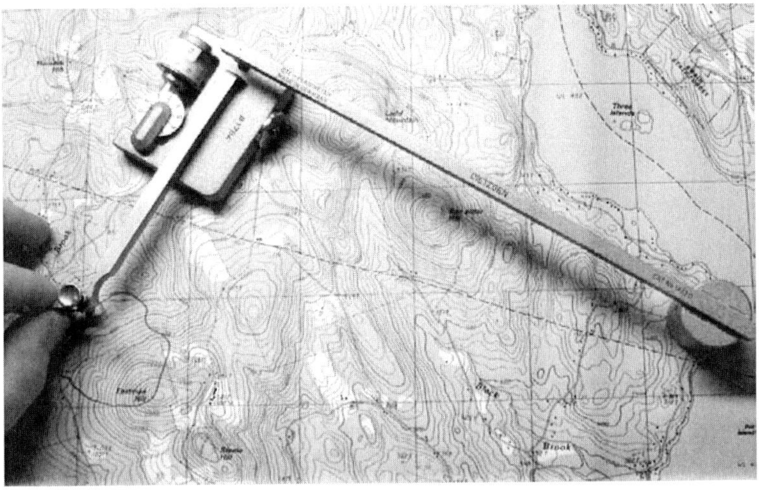

Fig. 2-17. Planimeter

Comparison

Another way to estimate the size of an area is by comparing it to areas of a known size. For example, if the area is entirely within a single section, then it can be safe to assume it isn't more than 640 acres; if it is the size of a football field, the estimate would be one acre.

Global Positioning System (GPS) Receiver

Several models of GPS receivers have the ability to calculate area (acreage).

Estimating Distances

Refer to the map scale for measuring distances on the map. On 7.5 minute quad topographic maps, the engineer's 20 scale ruler is a nice tool because 20 graduations on the ruler equals 1 inch which equals 2,000 feet. To calculate distances you will need a map wheel (Figure 2-18), string, paper with tick marks, or other measuring tool. A map wheel has a toothed wheel and as it moves it measures distances on maps. Set the map wheel's scale to match the map's scale and then roll the wheel along the route to be measured. Table 2-9 describes the steps to estimate distances using a map wheel.

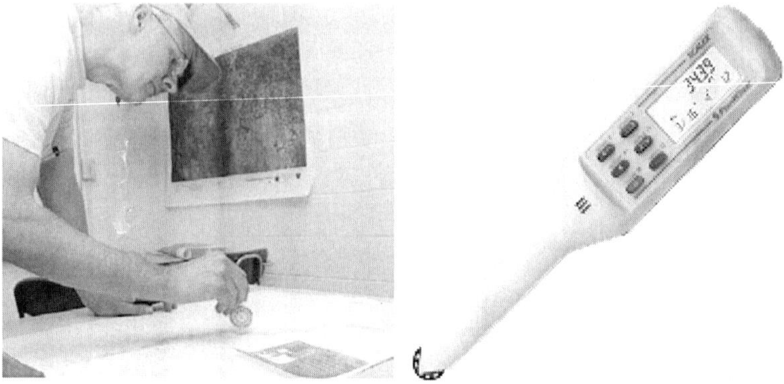

Fig.2-18. There are different types of map wheels.

Table 2-9. Steps for estimating distances using a map wheel.

Steps	Directions
1	Set map wheel to zero and calibrate using scale on the map.
2	Roll the map wheel along the distance to be estimated. Do this twice or more to get an average.

Estimating Per cent Contained

To estimate per cent contained, use this formula:

$$\frac{\text{Completed line distance}}{\text{Perimeter distance}} \times 100 = \text{Per cent contained}$$

Appendix IV

Procedure of Obtaining Aerial Photographs

For acquiring aerial photography, (either already existing or freshly flown) the users should write to Survey of India or National Remote Sensing Agency at following addresses:

a) Director

National Remote Sensing Agency

Balanagar, HYDERABAD - 500 037

b) Officer-in-charge

No.73 (APFPS) Party, Survey of India

West Block No.4, Wing No.4, 2nd Floor

R.K. Puram, NEW DELHI - 110 066

Satellite data products are available in the following types of formats :

1. High Density Digital Tape (HDDT)
2. Quick Look Film
3. Computer Compatible Tape (CCT), Digital Audio Tape(DAT) Compact Disc(CD-ROM)
4. 70mm film
5. 240 mm Black and White film positive/negative in individual band
6. Black and White paper prints & enlargement in individual band
7. 240 mm False Colour Composite (FCC) Film
8. FCC paper print and enlargements

The Earth Station situated at Shadnagar, Hyderabad tracks the LANDSAT, METSAT, ERS and IRS series of satellites. The data is primarily recorded on HDDT and Quick Look 70 mm format roll films. Standard products listed in (3) to (8) are generated from HDDTs and supplied to the users.

The NRSA Image Processing Laboratory located at NRSA, Balanagar, Hyderabad is the single repository of data supply in India on request to the users. The quality of the products, that leave NDC is checked in respect of computer processing, photographic reproduction, cloud cover and other defects as well. As an aid to the users for the selection of data, Browse Facility provides an array of computerized data listing, satellite coverage map, microfiche films and satellite orbital calendars.

All the user's data products requirements and inquiries can be addressed to :

Head,

NRSA Data Centre,

National Remote Sensing Agency,

Balanagar, Hyderabad - 500 037 (A.P.)

❏❏❏